Food Industry Briefing Se

OILS AND FATS IN THE FOOD INDUSTRY

Frank D. Gunstone

Professor Emeritus, University of St Andrews and Honorary Professor, Scottish Crop Research Institute, Dundee, UK

WILEY-BLACKWELL

A John Wiley & Sons, Ltd., Publication

Blackwell Publishing was acquired by John Wiley & Sons in February 2007. Blackwell's publishing programme has been merged with Wiley's global Scientific, Technical, and Medical business to form Wiley-Blackwell.

Registered office

John Wiley & Sons Ltd, The Atrium, Southern Gate, Chichester, West Sussex, PO19 8SQ, United Kingdom

Editorial offices

9600 Garsington Road, Oxford, OX4 2DQ, United Kingdom
2121 State Avenue, Ames, Iowa 50014–8300, USA

For details of our global editorial offices, for customer services and for information about how to apply for permission to reuse the copyright material in this book please see our website at www.wiley.com/wiley-blackwell.

Library of Congress Cataloging-in-Publication Data

Gunstone, F. D.
  Oils and fats in the food industry/Frank D. Gunstone.
    p.   cm. – (Food industry briefing series)
  Includes bibliographical references and index.
  ISBN-13: 978-1-4051-7121-2 (pbk. : alk. paper)
  ISBN-10: 1-4051-7121-9 (pbk. : alk. paper)
  1. Oils and fats. 2. Food industry and trade. I. Title.

TP680.G86 2008
664'.3–dc22
                      2008006198

A catalogue record for this book is available from the British Library.

Set in 10/13 pt FranklinGothic-Book by Charon Tec Ltd (A Macmillan Company), Chennai, India

Contents

Chapter 1   The Chemical Nature of Lipids   1

Chapter 1 describes those fatty acids that are important in foods and the glycerol esters in which these acids occur in the oils and fats used in the food industry. This is followed by a description of minor components also present including ester waxes, phospholipids, sterols, tocols, and hydrocarbons. Many of these minor components are valuable in their own right.

Chapter 2   The Major Sources of Oils and Fats   11

Chapter 2 is devoted to the major commercial sources of oils and fats. These are mainly of plant origin but there is still a significant use of animal fats. Selected sources are discussed in terms of production levels and composition.

Chapter 3   Extraction, Refining, and Modification Processes   **26**

Commodity oils have to be extracted from their agricultural source and are then usually refined to produce a bland product. However, the natural oils are not always optimum for their food-use purpose – they may fall short in terms of their physical, chemical, or nutritional properties – and procedures have been devised by which the oils come closer to what is desired. These procedures will be described. It is important that those in the food industry know what can be achieved and at what financial, environmental, or other cost.

Chapter 4   Analytical Parameters   **37**

Commodity oils and fats are generally purchased with a specification. This chapter describes the terms which might appear on a specification and the analytical procedures by which these are measured.

Chapter 5 Physical Properties   **59**

Important physical properties include crystallisation and melting, spectroscopic properties (covered in Chapter 4), and some others used in trading oils and fats.

Chapter 6 Chemical Properties   **71**

Fatty acids and their esters have a carboxyl group and frequently contain one or more unsaturated centres (double bonds). Each of these functional groups has characteristic properties and those of greatest importance in the food industry are reviewed.

OILS AND FATS IN THE FOOD INDUSTRY

Chapter 7   Nutritional Properties   89

It is obvious that the nutritional properties of fats and oils and their various components will be of interest to the food industry. There is increasing awareness of the link between diet and health/disease. Links between chemical structure and physiological properties are discussed first followed by recommended dietary intake and the role of lipids in some of the more important disease conditions.

Chapter 8   Major Edible Uses of Oils and Fats   113

The final chapter is devoted to a description of the most common food uses of oils and fats.

CONTENTS

Series Editor's Foreword

It was with some excitement that I received Professor Frank Gunstone's manuscript of this book, *Oils and Fats for the Food Industry*, which makes a very worthy addition to the *Food Industry Briefing Series*. I had already read two other books by Professor Gunstone and I felt that not only would I enjoy reading the manuscript for this book, but also that Professor Gunstone would add further to my knowledge of the subject and teach me something new. I was right on all counts.

With *Oils and Fats for the Food Industry* Professor Gunstone has proven once again his mastery of a subject matter of great importance to the food industry. It is written in a style that makes the concepts and information contained easily accessible. Importantly, the structure of the book is very logical, presenting the reader with a journey through the subject matter that will inevitably lead to increased knowledge and understanding. Significantly, this is a concise book. It does not attempt to answer all questions about oils and fats, and it does not attempt to stand in the place of weightier tomes. It fits precisely with the concept of the *Food Industry Briefing Series* as a food technology series that provides books which offer great utility to food industry professionals who wish to increase their knowledge of a given subject, but who have little time to devote to the task. With this book, those who wish quickly to gain insight to the topic of oils and fats used in the food industry can do just that. If they then wish to go on to read heavier volumes they will find that *Oils and Fats for the Food Industry* has provided the foundation on which to build their expertise.

The *Food Industry Briefing Series* was devised to provide the food industry with a resource that can be used by managers and executives to broaden their knowledge and gain expertise without devoting inordinate amounts of time to study. It was also conceived that the series would provide a resource for the personal development of staff whose career development is predicated on increasing their expertise. In addition to satisfying the needs of food industry professionals for quickly accessible texts on food technology topics, the *Food Industry Briefing Series* should also provide benefits to both lecturers in the fields of food science and technology and their students. Recognising also that food industry professionals, lecturers, their students, and university libraries are all subject to budgetary controls, the *Food Industry Briefing Series* has been conceived as a source of high-quality food technology texts that fall well below the price threshold of typical technical and academic texts.

Ralph Early
Series Editor, *Food Industry Briefing Series*
Harper Adams University College

SERIES EDITOR'S FOREWORD

Preface

The human diet contains three macronutients and several micronutrients (vitamins, minerals, antioxidants, etc.). The macronutrients are sources of different kinds of proteins, carbohydrates, and fats (lipids) and the food industry is concerned to supply these as primary products or as constituents of a wide range of foods. A healthy supply of macronutrients will generally contain the necessary micronutrients.

Despite the impression given by many uniformed sources that fat is an undesirable part of the diet, it remains an essential requirement. However, we are increasingly aware that both the quantity and the quality of the fat consumed are important elements of a healthy diet.

This book is an attempt to describe the nature of fat for those working in the food industry and for those in the media seeking to guide the public about what they should consume. It is impossible to do this without some reference to the chemical structure of fatty acids and lipids but structural features have been kept to the minimum. I have been generous in producing up-to-date numbers for production levels to give a better indication of the relative importance of both the starting materials and the products.

All the fats used by the food industry and consumed on a daily basis are products of the agricultural industry. Today they are mainly of plant origin and are grown on plantations or in fields in tropical and temperate regions. About one-sixth of the supply is still of animal origin but this underestimates human consumption because the fat eaten in meats and in dairy products (other than butter) is not monitored by commodity analysts. Some oilseeds and oily fruits are grown for their lipid content (palm, olive, rapeseed), others are co-products grown for oil and meal (soybean), and yet others are by-products of another industry (cottonseed oil and meat fats such as lard and tallow). Each oil has its own mixture of fatty acids and its own fingerprint of minor components. Where these are not optimum for human food use they may be subject to modification by biological procedures before planting or by technological procedures after extraction. These have been described.

While used mainly for human food fats have always found minor uses for animal feed and for a range of oleochemical industries using oils and fats as starting materials. Biodiesel falls into this last group. Because of concerns over several aspects of our traditional use of mineral oil there is a growing interest in biofuels. At present these consist mainly of bioethanol from appropriate carbohydrates and biodiesel from appropriate oils and fats. There is concern that these materials are being made from staple foods and the question is asked: will there be enough for both purpose? This matter is discussed, mainly in terms of the last 5 years and the next 5 years. Forecasts beyond that short-time frame are likely to be in error.

The lipids have important physical, chemical, and nutritional properties and these have to be brought into appropriate balance. This is not always an easy task. Nutritionists may indicate a recommended quantity and quality of fat and seed producers, farmers, and those in the agricultural and food businesses may strive to produce material to meet these targets. It remains only for the consumer to follow the advice. This is often a major difficulty. There are concerns at the present time with consumption levels of *trans* acids which need to be reduced and of omega-3 acids which need to be raised, and with the growing problem of obesity. This book is offered to those who wrestle with these problems in the hope that it might be of assistance.

To keep this volume short the number of references has been limited. Most of those cited are to reviews and books where further information may be obtained.

Frank D. Gunstone
St Andrews

Abbreviations and Acronyms

AA	Arachidonic acid
ACP	Acyl carrier protein
ALA	Alpha-linolenic acid
AMF	Anhydrous milk fat
CBE	Cocoa butter equivalent
CLA	Conjugated linoleic acid
DHA	Docosahexaenoic acid
EDTA	Ethylenediamine tetra-acetic acid
EPA	Eicosapentaenoic acid
ESR	Electron spin resonance
FTIR	Fourier transform infrared
GLA	Gamma-linolenic acid
HDL	High-density lipoprotein
IP	Identity preserved
L	Linoleic acid or ester
La	Lauric acid or ester
LCPUFA	Long-chain polyunsaturated fatty acids
LDL	Low-density lipoprotein
Ln	Linolenic acid or ester
Mt	Million tonnes
MUFA	Monounsaturated fatty acids
NIR	Near infrared
NMR	Nuclear magnetic resonance
O	Oleic acid or ester
PC	Phosphatidylcholine
PE	Phosphatidylethanolamine
PFAD	Palm fatty acid distillate
PI	Phosphatidylinositol

PMF	Palm mid-fraction
PS	Phosphatidylserine
PUFA	Polyunsaturated fatty acids
RBD	Refined, bleached, and deodorised
S	Stearic acid or ester
St	Saturated acids or esters
UV	Ultraviolet

Triacylglycerols are frequently represented by three letters such as POL. This symbol stands for all six of the triacylglycerols having palmitic, oleic, and linoleic acids attached to glycerol. Other three letter groupings are to be interpreted similarly.

ABBREVIATIONS AND ACRONYMS

The Chemical Nature of Lipids

The word 'lipid' is the scientific name given to a wide range of natural compounds based on fatty acids (or closely related molecules such as fatty alcohols or sphingosine bases). Oils and fats are an important subsection of lipids differing from one another in whether they are liquid or solid at ambient temperature. This physical property depends mainly on the fatty acids that they contain. Most animal fats are solids and most vegetable fats are liquid but there are some solid tropical vegetable fats often described as 'butters' (e.g. cocoa butter). This chapter is a review of the nature of fatty acids, triacylglycerols, and the lipid-soluble compounds that are present as minor but significant components of natural oils and fats.

1.1 Fatty acids

Well over 1000 natural fatty acids have been identified but most food scientists need to be familiar with only around 20 of these. Although there are exceptions, the most common fatty acids have a straight chain of 8–22 carbon atoms (even numbers only) and frequently have one or more unsaturated centres (almost entirely double bonds of *cis* configuration) occurring at preferred positions in the carbon chain.

The position of a double bond may be indicated by its relationship to the carboxyl group (delta or Δ nomenclature) or to the end methyl group (omega or ω nomenclature). The most common monounsaturated acid – oleic acid – is both $\Delta 9$ and $\omega 9$ (Figure 1.1). Acids are described as saturated, monounsaturated, or polyunsaturated (PUFA) depending on the number of unsaturated centres they contain. Double bonds in polyunsaturated acids are usually separated by one methylene group (CH_2). Consequently these acids contain structures such as the pentadiene unit shown below which have characteristic and important properties based on the presence of a doubly activated methylene group (see Chapter 6).

$$-CH=CHCH_2CH=CH-$$ the pentadiene unit present in polyunsaturated acids

Melting points (Chapter 5) increase with chain length and decrease with the number of (*cis*) double bonds. *Trans* compounds are higher melting than their *cis* isomers. These claims are illustrated in Table 1.1.

$$CH_3(CH_2)_7CH=CH(CH_2)_7COOH$$

Figure 1.1 Oleic acid [9c-18:1 or ω9-18:1]. The shorthand symbol indicates a C_{18} acid with one *cis* unsaturated centre starting on C-9 with respect to both the COOH and CH_3 groups.

Table 1.1 Names, structure, and melting points of the more common fatty acids. All the unsaturated acids listed here occur in the all-*cis* form

Trivial name	Systematic name	Short-hand	Double bonds*		MP (°C)	Molecular weight
			Δ	ω		
Butyric	Butanoic	4:0			−5.3	78.1
Caproic	Hexanoic	6:0			−3.2	116.2
Caprylic	Octanoic	8:0			16.5	144.2
Capric	Decanoic	10:0			31.6	172.3
Lauric	Dodecanoic	12:0			44.8	200.3
Myristic	Tetradecanoic	14:0			54.4	228.4
Palmitic	Hexadecanoic	16:0			62.9	256.4
Stearic	Octadecanoic	18:0			70.1	284.5
Oleic	Octadecenoic	18:1	9	9	16.2	282.5
Vaccenic	Octadecenoic	18:1	11t	7	44.1	282.5
Linoleic	Octadecadienoic	18:2	9,12	6	−5	280.4
Linolenic	Octadecatrienoic	18:3	9,12,15	3	−11	278.4
Arachidonic	Eicosatetraenoic	20:4	5,8,11,14	6	−49	304.5
EPA	Eicosapentaenoic	20:5	5,8,11,14,17	3		302.5
DHA	Docosahexaenoic	22:6	4,7,10,13,16,19	3		328.5

*When using ω nomenclature only one number is cited. This gives the position of the first double bond with respect to the end methyl group. It is assumed that additional double bonds display the usual methylene-interrupted pattern of unsaturation. For example, this term is commonly employed for the class of omega-3 acids. The term omega acids (without a number) has no meaning.

1.2 Triacylglycerols

Fatty acids occur mainly as glycerol esters. Glycerol contains three carbon atoms each of which carries a hydroxyl group [$HOCH_2CH(OH)CH_2OH$]. It can form several different types of esters. The outer hydroxyl groups on C-1 and C-3 differ from one another only in a subtle way (which will be largely ignored throughout this book) but the two primary hydroxyl groups differ more obviously from the secondary hydroxyl group attached to C-2. Depending on the number of hydroxyl groups that are acylated (*i.e.* esterified with a fatty acid or acyl group) the glycerol esters are termed monoacylglycerols (MAG), diacylglycerols (DAG), and triacylglycerols (TAG). These terms are preferred to the older names monoglycerides, diglycerides, and triglycerides though these latter are still frequently used in industry and in medical/nutritional texts.

There are two types of monoacylglycerol designated with numbers or Greek letters as for example 1-(or α-)monostearin and 2-(or β-)monostearin. The first of these forms the more stable crystals and an equilibrium mixture of the two will be 90% 1-monostearin. There are also two isomers of distearin designated as 1,2-distearin and 1,3-distearin. These are sometimes described as the symmetrical (1,3-) and unsymmetrical (1,2-) isomers. More frequently there will be two different acyl groups and this will increase the number of isomeric forms (Figure 1.2).

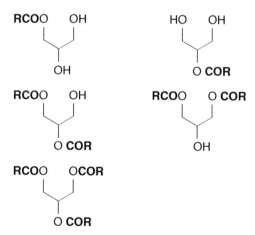

Figure 1.2 Glycerol esters (1- and 2-MAG, 1,2- and 1,3-DAG and TAG). **RCO** represents the acyl group from the fatty acid **RCOOH**. All other letters relate to atoms derived from the glycerol molecule.

When a triacylglycerol contains only one kind of fatty acid then there is only one form of this (*e.g.* triplamitin, PPP). With two different acids there will be a symmetrical and an unsymmetrical isomer such as POP and PPO where P and O represent palmitic and oleic acid in the 1, 2, and 3 positions. If a fat contains three different fatty acids then it can theoretically be a mixture of 18 different TAG containing one or more of these acids though in practice some of these will be present only at low levels. These are detailed in Table 1.2. Three-letter codes such as POP represent palmitic acid at positions 1 and 3 and oleic acid at position 2. Isomeric TAG containing P, O, and St exist as POSt, PStO, and OPSt (ignoring the subtle difference between POSt and StOP). However a fat containing these three acids may also contain other TAG as indicated in Table 1.2. Since most fats contain at least four or five different fatty acids the number of potential TAG is even larger, and a natural fat is usually a mixture of all possible structures. However the distribution of acyl groups between the glycerol hydroxyl groups is not generally random because of the regioselectively controlled reactions by which these molecules are assembled under the influence of enzymes. In vegetable fats particularly, the *sn*-2 position of glycerol tends to be acylated with unsaturated rather than with saturated acids so, for example, the proportion of POP (the symmetrical isomer) will greatly exceed that of PPO (the unsymmetrical isomer). Because of this non-random distribution the triacylglycerol composition of natural fats (and hence their physical and biological properties) can

Table 1.2 Eighteen TAG can be present in a fat containing only palmitic (P), oleic (O), and stearic (St) acids

Number of different fatty acids	TAG
One	PPP, OOO, StStSt
Two	PPO, POP, PPSt, PStP, OOP, OPO, OOSt, OStO, StStP, StPSt, StOSt, StStO
Three	POSt, OPSt, PStO

These three-letter structures are to be read as indicating the fatty acids present in the *sn*-1, 2, and 3 positions of glycerol, respectively. There are subtle differences between the *sn*-1 and 3 positions so that PPO differs only in special ways from its enantiomer OPP but this has been ignored in this table. Symbols such as POSt are used here to represent a single triacylglycerol molecule (actually a mixture of two enantiomers – POSt and StOP). Sometimes these three-letter symbols are used to represent all the different TAG that contain these three acids. The reader must be alert to decide which on the basis of the context.

be changed through randomisation of the acyl groups (interesterification). In the past lard was frequently interesterified because the randomised product (with modified triacylglycerol composition) was a better shortening than the non-randomised product.

1.3   Ester waxes

The term wax is used to describe materials with certain physical appearance and properties. They are generally mixtures of several types of medium- and long-chain compounds including hydrocarbons (RCH_3), alcohols (RCH_2OH), aldehydes (RCHO), acids (RCOOH), and esters (RCOOR'). The last of these are better classified as ester waxes with around 40 carbon atoms in each molecule formed mainly from saturated or monounsaturated long-chain fatty acids and fatty alcohols. Thus $RCO-OR'$ represents the ester from the acid $RCO-OH$ and the alcohol $R'O-H$. Waxes are of both vegetable (carnauba, jojoba) and animal (beeswax, woolwax) origin. Some solvent-extracted oils contain low levels of wax from the seed coatings (hulls) and may need to be dewaxed (Section 3.5).

1.4   Phospholipids

The phospholipids (often referred to as phosphatides, Figure 1.3) are molecules of the highest biological importance. Every cell in living plants or animals is a sac within which essential life-supporting processes take place. This sac is surrounded by a permeable membrane made up of phospholipids and other lipid components. The membrane permits the controlled movement of molecules into and out of the cell. These movements are dependant on the quality and integrity of the cell membranes. Phospholipids also have significant properties based on their amphiphilic nature that make them important in foods, cosmetics, and pharmaceuticals. They are largely removed during the refining of vegetable oils at the degumming stage but can be recovered as lecithin. This is a crude mixture of different phospholipids (see Figure 1.3 for typical structures) along with TAG and glycolipids. Lecithin can be used in this crude form but more usually it is refined and is supplied as a phospholipid concentrate virtually free of TAG

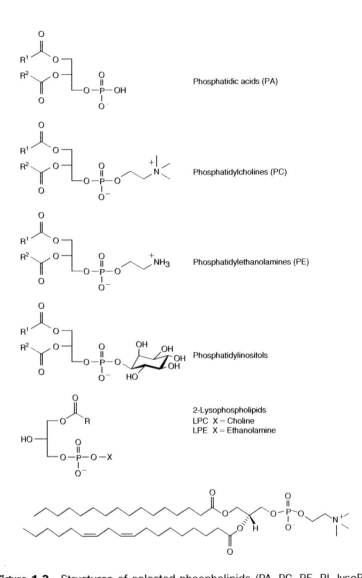

Figure 1.3 Structures of selected phospholipids (PA, PC, PE, PI, lysoPL). These are correctly named in the plural because natural products are mixtures of compounds varying in the nature of the acyl groups R^1CO and R^2CO. Also an alternative representation of a PC containing palmitic acid and linoleic acid is shown. These molecules (apart from phosphatidic acid) contain four ester bonds. On complete hydrolysis they furnish fatty acids, glycerol, phosphoric acid, and a hydroxy compound (choline, etc). A series of phospholipases exist which catalyse selective hydrolysis (lipolysis) of these ester groups. Most of these structures have been taken from 'Lipid Glossary 2' (The Oily Press, 2000) which can be downloaded free via The Oily Press website by permission of the authors and the publisher.

or as individual phospholipid classes (such as phosphatidylcholines). The phospholipids can be further modified by chemical or enzymatic processes to make them more suitable for a particular purpose.

Crude soya bean oil (~3%), rapeseed oil (~3%), sunflower oil (~1%), and palm oil (negligible) contain phospholipids at the levels indicated and the first three (particularly soya bean) are the major source of commercial lecithin. A typical crude soya bean lecithin contains phosphatidylcholines (PC, 10–15%), phosphatidylethanolamines (PE, 9–12%), and phosphatidylinositols (PI, 8–10%) along with other lipids.

When hydrolysed, glycerophospholipid molecules give glycerol, fatty acids (2 mols), and low molecular weight alcohols frequently containing nitrogen as in choline ($HOCH_2CH_2NMe_3$) and ethanolamine ($HOCH_2CH_2NH_2$). Individual classes of phospholipids are still mixtures because of variation in the fatty acids present. Some typical structures are given in Figure 1.3.

1.5   Sterols and sterol esters

Though not strictly lipids, sterols occur in many oils and fats, share some of their physical properties, and are closely related to oils and fats in any discussion on food and health. Cholesterol has long been associated with coronary heart disease (CHD) and many recommendations concerning lipid intake are related to its influence on the cardiovascular system (Section 7.13) (Figure 1.4).

Most crude vegetable oils contain 1000–5000 mg/kg (0.1–0.5%) of phytosterols, partly as free sterols and partly as sterols esterified with fatty acids. The ratio of esterified to free sterol varies with free sterols (40–80%) generally predominating. Palm oil and the two lauric oils have very low levels of phytosterols (40–50 mg/kg). Above average levels are present in rapeseed oil (5–11 g/kg, mean ~7.5) and in corn oil (8–22 g/kg, mean 14). Sitosterol is generally the major phytosterol (50–80% of total sterols) with campesterol, stigmasterol, and Δ^5-avenasterol frequently attaining significant levels. These are present at lower levels in refined oils with some, but not all, having being partially removed in the deodorisation process. Sterols can be recovered from deodoriser distillate (along with other compounds) and these provide a valuable source of precursors of many pharmaceutical steroids. Phytosterols are now added to spreads and other foods to reduce cholesterol absorption (Sections 7.7 and 7.13).

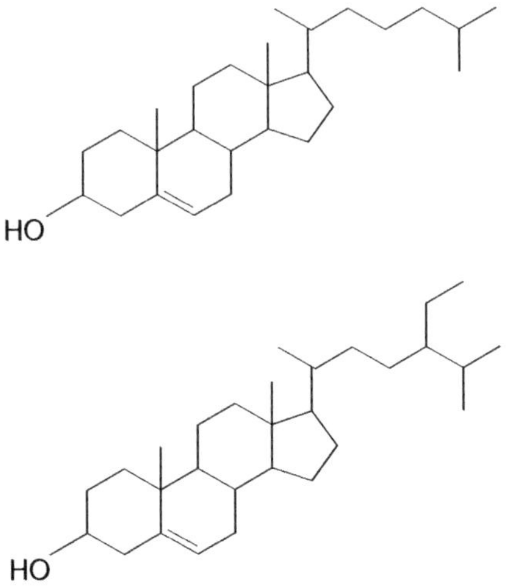

Figure 1.4 Cholesterol (upper) and sitosterol (lower).

Cholesterol is a zoosterol almost entirely of animal origin and is not present in plant systems at any significant level. The normal value of 20–50 ppm in vegetable oils compares with the much higher levels reported for depot animal fats (up to 1000 ppm), fish oils (up to 7000 ppm), dairy fats (2000–3000 ppm), and egg yolks (12,500 ppm). To advertise a vegetable oil as cholesterol-free is true but slightly misleading since such oils would not be expected to contain cholesterol.

1.6 Tocols

There are eight tocols (tocopherols and tocotrienols) with the structures shown in Figure 1.5. They are phenolic heterocyclic compounds with a C_{16} lipid-soluble isoprenoid side chain. The tocols have two valuable but not identical properties: they show vitamin E activity and they are powerful, natural lipid-soluble antioxidants. The total tocol content of crude vegetable oils can range from 10 ppm (equivalent to mg/kg) in coconut oil to 1370 ppm in soya bean oil. Levels of the tocopherols in selected samples of crude vegetable oils are listed in Table 1.3. The tocotrienols are insignificant in

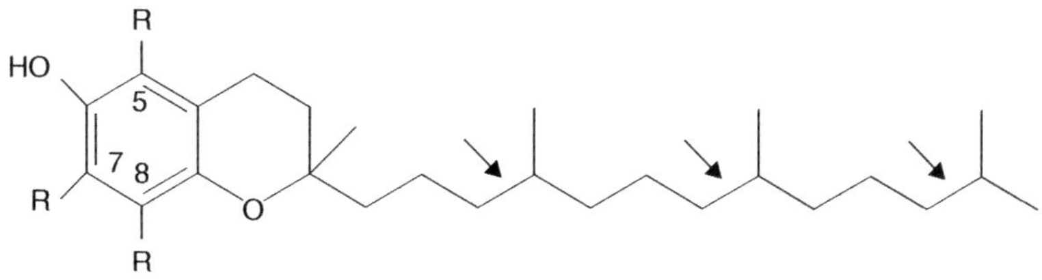

Figure 1.5 Tocopherols and tocotrienols. Tocopherols have a saturated C_{16} side chain, tocotrienols have double bonds at the three positions indicated by the arrows. R = H or CH_3; α = 5,7,8-trimethyltocol; β = 5,8-dimethyltocol; γ = 7,8-dimethyltocol; δ = 8-methyltocol.

Table 1.3 Levels (ppm) of the four tocopherols in crude rapeseed, palm, soya bean, and sunflower oils

	α	β	γ	δ
Rapeseed	179	0	415	10
Palm	190	0	0	0
Soya bean	120	10	610	190
Sunflower	610	10	30	10

Adapted from Warner (2007) *Lipid Technol.*, **19**, 231.

most oils other than palm oil and rice bran oil. Some tocopherols appear in the deodoriser distillate after refining and soya bean deodoriser distillate is the major source of mixed natural tocopherols, apart from the oils themselves. Tocols rich in tocotrienols can be recovered from palm oil. α-Tocopherol can also be made synthetically and this represents 80–90% of total available tocopherol.

Natural tocopherol mixtures are used as antioxidants at levels up to 500 ppm, frequently along with ascorbyl palmitate which extends the tocol antioxidant activity. At higher levels (>1000 ppm) α-tocopherol is considered to act as a pro-oxidant (Section 6.2). Since vegetable oils contain tocols at 200–800 ppm further additions show only a limited effect. The tocols themselves are very sensitive to oxidation and are more stable in esterified form where the all-important hydroxyl group is not free. However such esterified compounds do not show antioxidant activity until they have been hydrolysed *in vivo* to the free phenolic form.

1.7 Hydrocarbons

Oils and fats sometimes contain low levels of hydrocarbons. These include alkanes, squalene (present in some marine oils and at lower levels in olive and amaranthus oils) (Figure 1.6), and carotenes, and may be contaminated with polycyclic aromatic hydrocarbons (PAH). β-Carotene is the biological precursor of vitamin A (Figure 1.7).

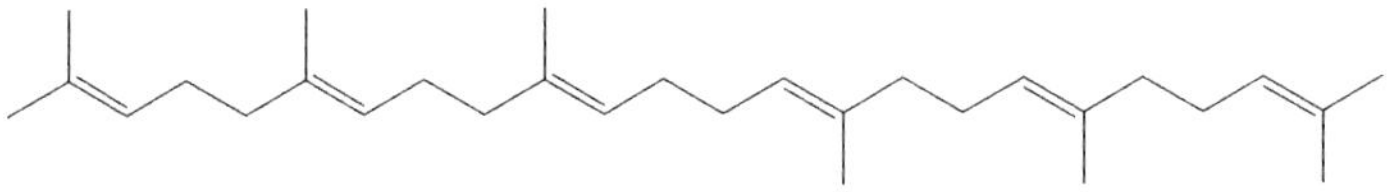

Figure 1.6   Squalene ($C_{30}H_{50}$).

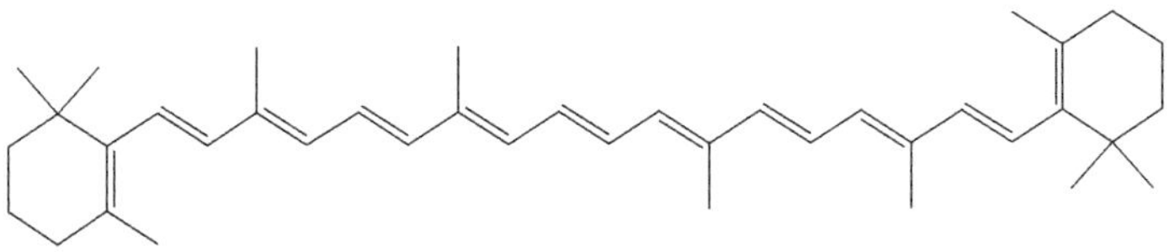

Figure 1.7   β-Carotene ($C_{40}H_{56}$). Other carotenes vary in the nature of the cyclic end groups.

The Major Sources of Oils and Fats

2.1 Introduction

One market analyst reports that in the harvest year 2006/07 (*i.e.* harvests from the Northern hemisphere in 2006 and from the Southern hemisphere in 2007) 153 million tonnes of oils and fats were produced from 17 commodity sources. These comprise four major oils (palm, sunflower, rape or canola, and sunflower), four minor oils (cottonseed, groundnut or peanut, olive, and corn), two lauric oils (palm kernel and coconut are rich in lauric acid – 12:0), three oils produced at only low levels (sesame, linseed, and castor), and four animal fats (tallow, lard, butter, and fish oil). These are listed in that order in Table 2.1. It is clear that supplies are dominated by four vegetable oils (palm, soybean, rapeseed, and sunflower) and increasingly by palm oil (from Malaysia and Indonesia) and soybean oil (from Argentina, Brazil, and USA).

A part of this total supply is used by the oleochemical industry as starting material for a range of industrial products. This holds for all the castor oil, most of the linseed oil, a significant proportion of the lauric oils and of tallow, and some of the remaining oils.

Table 2.1 Production (million tonnes) of 17 commodity oils and fats in 2006/07

Palm	38.0	Olive	3.0	Castor	0.5
Soybean	36.7	Corn	2.3	Tallow	8.5
Rapeseed	18.1	Palm kernel	4.4	Lard	8.0
Sunflower	11.4	Coconut	2.9	Butter	6.9
Cottonseed	5.0	Sesame	0.9	Fish oil	1.0
Groundnut	4.2	Linseed	0.7	*Total*	152.7

Note: Forecasts for 2007/08 are for 160.4 million tonnes.
Source: Adapted from Oil World Annual (2007), ISTA Mielke GmbH, Hamburg.

Some is also used for animal feed but the rest of these commodity oils form the basis of the food industry. For many years it has been assumed that these 17 oils and fats were used for food, feed, and the oleochemical industry in the ratio 80:6:14 but this is changing, mainly through the rapidly growing demand for biodiesel. At the end of 2007 it is probably close to 74:6:20. It has been suggested that by 2020 it could be 68:6:26 though recent evidence suggests that the shift may be greater than this. This revised ratio does not mean that less will be consumed as food but that this will represent a smaller proportion of a larger total. In 2007/08 one half of the year's additional supply will probably be used for non-food purposes (mainly biodiesel) and the other half for food purposes. During 2007 and perhaps for some time to come demand exceeds supply and demand will be restricted through price rationing.

It is worth noting that this arbitrary list of commodity oils does not include cocoa butter nor many minor oils now available on the market such as camelina oil, almond oil, and evening primrose and borage oils to mention only a few. In dietary terms further significant quantities of fat are ingested which do not appear in the listings of oil and fat market analysts. For example, dairy produce (other than butter) such as milk, cream, and cheese and fats ingested when meat or fish is consumed or when nuts are eaten.

With the burgeoning demand for oils and fats as a source of biodiesel, concern has been expressed that there will not be enough for food purposes as the demand for this also increases with growth in population and income. It is known that with increasing income, particularly in developing countries, there is a growing demand for meat and fat. Total production of oils and fats is likely to increase through harvesting larger areas and through improved yields. Nevertheless, oil and fat prices are already under pressure, and this will influence the demand for both food and non-food purposes until a new equilibrium is reached.

After a period of below-trend prices about 10 years ago, reaching a nadir in 1997/98, prices of oilseeds and of oils and fats have risen very sharply in recent years. This reflects enhanced demand both for food purposes, especially in China and other developing countries and for conversion to biodiesel. At the same time there is a tightened supply situation arising from unfavourable weather conditions in many producing countries.

Over the past 10 years (1997–2006) annual production of oils and fats has risen on average by around five million tonnes. Over the last 3 years the figure has been close to eight. It has been estimated that in the next 5 years (2008 onwards) demand for food and for non-food purposes could each rise by about 4–5 million tonnes. Can supplies increase by 8–10 million tonnes in the near future to meet the demand for food and fuel? The figures cited above allow for a pessimistic or an optimistic conclusion. The closeness of the supply and demand figures will operate a rationing system through stocks and prices. Of course other factors will be relevant, in particular the supply/demand/price situation for wheat and other cereals and the cost of (mineral) oil.

The question of vegetable oils derived from genetically modified seeds is still an issue in Europe and in some other countries. Over one half (and increasing) of soybean oil comes from genetically modified seeds in both North and South America and also increasing proportions of canola oil from Canada. Non-GM material must be sourced from appropriate areas and be identity preserved (IP). The first GM oilseeds were modified to be resistant to certain herbicides and to pests leading to more efficient farming. The second generation of GM oilseeds have been modified to have different fatty acid composition and/or enhanced levels of minor components. There may be pressure for Europeans to accept oil from GM plants for industrial (biodiesel) purposes.

When oilseeds are extracted by pressing and/or use of solvent (Section 3.1) they furnish differing proportions of oil (lipid) and meal, rich in protein and used mainly for animal feed. The lipids from the different vegetable and animal sources vary in fatty acid composition. They consist mainly (around 95%) of triacylglycerols based on component acids which are chiefly palmitic (16:0), oleic (18:1), and linoleic (18:2) with other acids such as stearic (18:0) and linolenic (18:3) at lower levels and yet other acids occurring only occasionally. The native oils also contain free acids, and mono- and diacylglycerols (at low levels), phospholipids (up to 3%), free sterols and sterol esters (0.1–2.2%), tocopherols (up to 0.2%), and other minor components (Chapter 1). After refining, the triacylglycerol content is usually above 99% and the other components are correspondingly reduced. Several of the minor commodities have potential value and may be recovered from side streams in the refining processes. Sometimes they are added back into the oil or meal or into a food.

Some of the more important oils are discussed in the following sections and are presented in alphabetical order.

2.2 Animal fats (butter, lard, tallow, chicken fat, and fish oils)

(Cow) milk fat is a significant product of the agricultural industry and is consumed as milk, cream, butter, or cheese. Data on the annual production of butterfat is given in Table 2.2. Butter has some disadvantages (see below) but it is appreciated for its superior flavour. It is used mainly as a spread but also as a baking and frying fat. Its fatty acid composition is different from that of any other commodity oil with over 500 components already identified. These are mainly saturated acids (4:0 to 18:0, 65–70% wt) with some 18:1 which is mainly oleic acid but also vaccenic acid (11t-18:1) and other *trans* isomers resulting from biohydrogenation within the rumen. The content of PUFA is low (~3%). Also present at low levels are conjugated linoleic acids (CLA, mainly rumenic acid 9c11t-18:2), branched-chain acids, oxo (keto) acids, hydroxy acids, and lactones some of which are responsible for the characteristic flavour of butter. The composition of milk fat depends on the cows' diet and in some parts of the world this leads a slightly different composition in the summer (pasture-fed) and winter (fed indoors). The disadvantages of butter are considered to be its high fat level (80–82% – required by legal definition) in comparison with the lower levels now present in most spreads, its high level of saturated acids, its content of cholesterol, and the fact that it does not spread easily when taken directly from

Table 2.2 Typical fatty acid composition (%wt) of major animal fats

	14:0	16:0	16:1	18:0	18:1[a]	18:2	Other
Butter[b]	12	26	3	11	28	2	18
Lard	2	26	5	11	44	11	1
Beef tallow	3	27	11	7	48	2	2
Mutton tallow	6	27	2	32	31	2	0
Chicken fat	1	22	6	7	40	20	4

Notes
[a]Including *trans* isomers.
[b]Also 4:0 (3%), 6:0 (2%), 8:0 (1%), 10:0 (3%), and 12:0 (4%).
These figures are cited as wt%. Different figures result when the results are given on a molar basis, particularly when the fatty acids have a wide range of molecular weight as in butter. In a typical example the figures for butterfat were given as: 4:0 (3.7% wt and 9.6% mol), 6:0 (2.4% and 4.8%), 8:0 (1.4% and 2.2%), 10:0 (2.9% and 3.9%), 18:0 (13.9% and 11.4%), and *cis*-18:1 (28.0% and 23.0%).

the refrigerator. Some of these disadvantages have been circumvented but since butter is defined legally some of the modified products cannot be described as 'butter'. Butter contains CLA and other minor components (e.g. sphingolipids) for which nutritional benefits have been claimed. Anhydrous butterfat can be fractionated to give higher and lower melting fractions which find specific uses in the baking industry. In the Indian sub-continent milk fat is converted to ghee. This is a butter-like product with more fat and less water ($<0.2\%$) that keeps better in hot countries.

Lard is the fat produced by pigs and tallow is the fat from cattle (and sheep). The fatty acid composition of lard and tallow differ mainly because tallow comes from ruminants and is a harder fat because of biohydrogenation. Both fats contain cholesterol (3000–4000 mg/kg) and are deficient in polyunsaturated acids and in antioxidants. Only the highest grades of tallow can be used for food purposes. Lower grades are used as animal feed, for surfactants, and as a source of biodiesel. Lard is unusual among fats in that much of its palmitic acid is attached to the sn-2 position in glycerol. The only other important fat displaying this property is human milk fat. Because of this unusual distribution of fatty acids lard has changed physical properties after interesterification and is thereby converted to an improved shortening.

Despite the high level of poultry meat now consumed only limited quantities of chicken fat are available. This comes from food companies processing and cooking chickens on an industrial scale.

Fish oils differ from vegetable oils and from other animal fats in the wider range of chain length of their component fatty acids (C_{14} to C_{24}) including the very important omega-3 acids such as eicosapentaenoic acid (EPA 20:5) and docosahexaenoic acid (DHA 22:6). Fish oil is now a by-product in the production of fish meal and both meal and oil are used mainly in the diets of farmed fish. The human need for omega-3 LC-PUFA is met in part by eating fish and in part by consuming high-quality fish oil. This last may be taken in capsules, be incorporated into animal diets to enhance the levels of EPA and DHA in milk or eggs or flesh, or be added to prepared foods such as spreads, bread, and fruit drinks, often as a free-flowing microencapsulated powder. Encapsulation also provides some protection from oxidation (Williams and Buttriss, 2006). Fish oils enriched in EPA and/or DHA are available, usually in encapsulated form, as food additives or as dietary supplements.

2.3   Cocoa butter and cocoa butter alternatives

The cocoa bean provides cocoa powder and a solid fat (cocoa butter) both of which are important ingredients of chocolate. Cocoa butter (mp 32–35°C) is characterised by its high content (70–90%) of triacylglycerols of the type SOS where S stands for any saturated acid. Cocoa butter is rich in POP, POSt, and StOSt as a consequence of the presence of near-equal quantities of palmitic, stearic, and oleic acids (Table 2.3). Glycerol esters of this type are hard and brittle at ambient temperatures giving chocolate its characteristic snap and its steep melting curve with virtually complete melting at mouth temperature. This provides a cooling sensation and a smooth creamy texture. See also Sections 5.1 and 8.7.

Because cocoa butter carries a premium price cheaper fats with similar physical properties have been produced. In the EU cocoa butter can be replaced only in part by one or more solid vegetable fats confined to a prescribed list if the name chocolate is to be retained. These materials are described as cocoa butter equivalents (CBE). European chocolate may contain around 25–35% of cocoa butter and up to 5% (maximum) of the fats listed in Table 2.3. The only processes allowed in the production of these fats are refining and fractionation. This excludes hydrogenation and interesterification even when the latter is enzymatic. If fats outside this

Table 2.3 Major triacylglycerols in cocoa butter and in tropical fats that may partially replace cocoa butter in chocolate according to the EU Chocolate Directive

Common name	Botanical name	Major triacylglycerols (%wt)		
		POP	POSt	StOSt
Cocoa butter	*Theobroma cacoa*	16	38	23
Palm mid-fraction	*Elaies guinensis*	57	11	2
Borneo tallow (Illipe)	*Shorea stenoptera*	6	37	49
Kokum butter	*Garcinia indica*	1	5	76
Mango kernel stearin	*Mangifera indica*	2	13	55
Sal stearin	*Shorea robusta*	1	10	57
Shea stearin	*Butyrospermum parkii*	1	7	71

Note: The chocolate directive, agreed after about 30 years of negotiation, is based on political compromise rather than technical necessity and there are other regions of the world with a more liberal attitude as to what constitutes chocolate.

prescribed list with similar melting properties are used the confections cannot be described as 'chocolate'. Cocoa butter replacers (CBR) and cocoa butter substitutes (CBS) are generally lauric fats or vegetable oils (soybean, cottonseed, palm) that have been fractionated and partially hydrogenated (thereby producing higher melting *trans* acids). These materials can be used to make confectionery fats but products containing them cannot be described as chocolate in Europe. The EU rules represent a political-economic rather than a technical compromise and do not apply on a global scale. Cocoa beans yield more cocoa powder than is needed for the cocoa butter also produced. It is therefore desirable to find other uses for cocoa powder or to extend supplies of cocoa butter with alternative fats having the required physical properties.

2.4   Lauric oils (coconut, palm kernel)

Coconut and palm kernel oil (Tables 2.1 and 2.4) differ from other commodity oils in that they are rich in medium-chain saturated acids (8:0 to 14:0, especially lauric acid – 12:0) and have correspondingly less oleic and linoleic acid. They have iodine values of only 7–10 (coconut oil) and 14–21 (palm kernel oil). They are extensively used to produce surface-active compounds in addition to their wide use in the food industry as components of spreads, of non-dairy creams and coffer whiteners, and as alternatives to cocoa butter. To extend their range of usefulness they may be fractionated and/or hydrogenated. After hydrogenation both oils have iodine values around 2 and slip melting points of 32–34°C. The C_8 and C_{10} acids can be separated and reconverted to triacylglycerols. These 'medium-chain triglycerides' (MCT) are used as safe lubricants in food-making equipment or as easily metabolised fats for invalids and athletes.

The triacylglycerols of these lauric oils are usually reported by carbon number which is the sum of the carbon atoms in the three

Table 2.4 Typical fatty acid composition (%wt) of coconut and palm kernel oils

	8:0	10:0	12:0	14:0	16:0	18:0	18:1	18:2
Coconut	8	7	48	16	9	2	7	2
Palm kernel	3	4	45	18	9	3	15	2

acyl groups. They range from 28 to 54 and are predominantly in the range 32–44. The C_{36} group will be mainly, but not entirely, trilaurin (12,12,12).

2.5   Olive oil

Olive oil comes almost entirely from Mediterranean countries. It has a high reputation as a healthy oil based on its fatty acid composition, its minor components, and its method of extraction. It is available in several grades tightly specified by several EU directives that relate to its isolation procedure and subsequent treatment. Oil is obtained from the olive mesocarp (soft fleshy fruit) by pressing. The first pressings are of highest quality and are designated as virgin oils. This extraction procedure requires no solvent and avoids high temperature. The oil is not refined and so retains important minor components that add to its nutritional value.

There are nutritional and technical reasons why, for many purposes, oils should be low in saturated and polyunsaturated acids and high in oleic (or other monounsaturated) acid. This is one of the reasons for the status of olive oil (Table 2.5). Because of the high level of oleic acid olive oil consist mainly of triacylglycerols with two or three oleic chains (OOO 40–59%, POO 12–20%, and LOO 12–20%). The importance of this type of fatty acid composition is reflected in the fact that high-oleic varieties have been developed for most of the commodity oils such as soybean, rape, sunflower, and safflower (see following sections and Table 2.5). Sometimes these have been obtained through conventional seed breeding: sometimes they have resulted from genetic modification. Even those seeds whose fatty acid composition may have been modified by conventional seed breeding may also be genetically modified to add desirable agricultural traits. Until genetically modified crops are more widely accepted it is necessary to know the detailed origin of vegetable oils used in the food industry.

A number of minor oils also fall into the category of oleic-rich such as almond oil with 65–70% oleic acid, hazelnut (74–80%), macadamia (55–65% along with 16–25% of 16:1) and moringa oil (77%). Hazelnut is sometimes added to olive oil as an adulterant, exploiting the very similar fatty acid composition of these two oils. Rapeseed/canola oil (Section 2.7) is also an oleic-rich oil.

Table 2.5 Typical fatty acid composition (%wt) of selected vegetable oils

Oil	Major acids			Other acids		
	Palmitic	Oleic	Linoleic	Stearic	Linolenic	Other
Palm						
Oil	44	39	11	4		
Olein	41	41	12	4		
Stearin	47–74	16–37	3–10	4–6		
Mid-fraction	41–55	32–41	3–11	5–7		
Soybean						
Oil	11	23	54	4	8	
Low-len	10	41	41	5	2	
High oleic	6	86	2	1	2	
High sat	24	9	38	19	10	
Oil IV 132[a]	11	22	55	4	7	
Oil IV 110[a]	10	42	40	4		9[b]
Oil IV 97[a]	13	48	30	6		13[b]
Oil IV 81[a]	11	73	11	5		32[b]
Oil IV 65[a]	11	75	–	14		40[b]
Rapeseed/canola						
High erucic	4	15	14	1	9	20:1 10 22:1 45
Low erucic	4	62	22	1	10	
High oleic	3	78	10	3	3	
Low linolenic	4	61	28	1	2	
Sunflower						
Regular	11–13[c]	20–30	60–70			
High oleic	9–10[c]	80–90	5–9			
Mid-oleic (Nu-sun)	<10[c]	55–75	15–35			

Notes
[a] Soybean of IV 132 and after hydrogenation to the iodine value indicated.
[b] Content of *trans* acids.
[c] Total saturated acids.
Similar data are given in Table 8.4.

2.6   Palm oil

The oil palm produces two different oils (palm oil and palm kernel oil) in a ratio of around 8.5:1. Palm oil is now produced in larger quantities than any other fatty oil (Table 2.1, 38.0 million tonnes in 2006/07) and is even more dominant among traded commodity oils. It is grown mainly in Malaysia and Indonesia and is exported from these countries (total exports 30.4 million tonnes) to virtually all oil-importing nations and particularly to China (5.7 million tonnes), EU-27 (4.9 million tonnes), and the Indian sub-continent (India, Pakistan, and Bangladesh together 6.1 million tonnes). Imports to the USA are small (0.71 million tonnes) but have increased in recent years following attempts to reduce the levels of *trans* acids in processed fats. The oil palm produces more lipid/hectare than any commodity oilseed. Palm oil is exported as crude oil, as refined oil, and as fractionated palm olein and palm stearin. Red palm oil is deodorised at below 150°C so that it retains about 80% of its original carotene and is used as a dietary source of important compounds required to combat blindness. β-Carotene is the biological precursor of vitamin A. Palm oil is generally the cheapest of the commodity oils and is produced entirely without genetic modification and without solvent extraction. The major export from Malaysia is RBD (refined, bleached, and deodorised) palm olein and it may be necessary to check whether the oil being used is palm oil or palm olein.

The range of use of palm oil is extended through fractionation (Section 3.5). In its simplest form this affords stearin, olein, and mid-fractions but the process has been improved and extended to give sub-fractions ranging in iodine value from 17 to 72 compared with a value of 51–53 for palm oil itself. Palm olein (IV 56–59) is used mainly as a frying oil, palm stearin (IV 40–42) as hardstock, and palm mid-fraction (range of iodine values between 32 and 47) in confectionery fats.

Palm oil contains roughly equal amounts of saturated and unsaturated acids (Table 2.5). Depending on the presence or absence of palmitic acid palm oil triacylglycerols have 48–54 carbon atoms (excluding the three glycerol carbon atoms) and are mainly C_{50} (16, 16, and 18) and C_{52} (16, 18, and 18). The major triacylglycerol species are POP (29%), POO (23%), PLO (10%), and PLP (10%) (average values from a large survey of Malaysian samples). Palm oil contains ~5% of diacylglycerols. The crude oil is low in phospholipids (5–130 ppm) and contains carotenes (500–700 ppm), sterols (200–600 ppm), and tocopherols and tocotrienols (together 700–1100 ppm).

2.7 Rapeseed (canola) oil

Traditionally rapeseed oil (known formerly as colza oil) was characterised by high levels of erucic acid (22:1) and high-erucic oils (HEAR) still find use as a source of the valuable erucamide (RCONH$_2$) for use in plastic film. For food purposes HEAR has been replaced almost entirely by a low-erucic variety developed first in Canada and designated canola oil. This oil contains only low levels of erucic acid in the oil (<2%) and low levels of glucosinolates in the meal (<30 µM/g). The high demand for rapeseed oil for food purposes is now augmented, in Europe particularly, by its use as a source of biodiesel in the form of rapeseed oil methyl esters. It is a high-quality food oil with the lowest level of saturated acids among all commodity oils, a high level of oleic acid, and useful levels of linoleic and linolenic acid in a very good ratio (Table 2.5). However, the presence of linolenic acid means that for frying and some other purposes the oil will be subject to brush hydrogenation (Section 3.6) to reduce the level of this triene acid before use. The major triacylglycerols are LOO (23%), OOO (22%), LnOO (10%), LLO (9%), and LnLO (8%). The rapeseed plant lends itself to modification by traditional seed breeding or by genetic modification and many potentially valuable modified oils have been reported including high-oleic, low-linolenic, and other mutants.

2.8 Soybean oil

Soybeans are grown more than any other oilseed, mainly in USA, Argentina, Brazil, and China. When extracted, the beans furnish oil (18%) and a high-quality protein meal (79%). The latter is extensively used as animal feed and also in many processed human foods. Most of the soybeans now grown are genetically modified (mainly for agricultural reasons) and non-GM beans (or products) are only available as IP material. Soybean oil (36.7 million tonnes in 2006/07) is now second to palm oil in annual production. It is an unsaturated oil rich in linoleic acid (54%) and containing some linolenic acid (11%). As a consequence of this fatty acid composition around one half of the triacylglycerol molecules contain two or three linoleic acid chains. Typical values are LLO 16%, LLL 16%, LLP 12%, LLLn 7%, and LLSt 3%. Other triacylglycerols at levels exceeding 5% include LOP 9%, LOO 8%, and LOLn 6%. These three letter symbols include all isomeric triacylglycerols with the acyl chains indicated. Linolenic acid is easily oxidised (Section 6.2) thereby reducing the shelf life of foods containing this oil. This difficulty is reduced by 'brush hydrogenation' – a very light hydrogenation in which the level of linolenic acid is halved (Sections 3.3 and 6.1). An alternative is to use soybean varieties (some of which are non-GM) with lower levels of this acid (3–4%) that are now increasingly available, at least in North America (Table 2.5). Subjected to further hydrogenation regular soybean oil furnishes products suitable for spreads and cooking fats. Concerns about the undesirable effects of *trans* acid that are known to increase LDL and reduce HDL levels have led to the US requirement (perhaps to be followed elsewhere) for the labelling of *trans* content. Since most of the *trans* acids are formed during partial hydrogenation of PUFA attempts have been made to use alternative technologies such as interesterification (Section 3.7) to avoid or limit partial hydrogenation without at the same time raising the content of saturated acids (Table 2.5). Because of its dependence on soybean oil the problem of *trans* acids formed during partial hydrogenation is greater in the USA than in Europe. Crude soybean oil contains several minor components. Insofar as these are removed during refining they can be recovered and serve as valuable starting points for other purposes. Degumming removes most of the phospholipids (~3%) and gives lecithin which is a crude mixture of phospholipids with triacylglycerols and glycolipids (Section 1.4). Deodorisation produces soybean deodoriser distillate enriched in sterols (~18% from an original level of 0.33%) and tocols (~11% from an original level of 0.15–0.21%) and a valuable source of phytosterols and of vitamin E. Attempts are being made to produce soybean oils with changed fatty acid composition or with different levels of minor components. Some of these are detailed in Table 2.5.

2.9   Sunflower seed oil

Sunflower seed oil is the last of the four major vegetable oils. In addition to its regular high-linoleic variety it is available in high-oleic and mid-oleic varieties. None of these are GM-crops. The commodity oil has been much favoured in Europe as a constituent of soft margarines free of linolenic acid. However, this advantage could be viewed differently as the need for additional omega-3 acids in the human diet is accepted. Details are given in Table 2.5. The major triacylglycerols in the regular oil are typically LLO (30%), LLL (27%), LOO (11%), LLP (10%), and PLO/StLL (10%). Sunflower oil may contain a wax (300–600 ppm) from the seed hulls unless the oil has been dewaxed (Section 3.5).

2.10   Other vegetable oils

Supplies of vegetable oils are dominated by palm oil and soybean oil followed by rapeseed/canola and sunflower oils. These, along with olive oil, cocoa butter, and the two lauric oils (coconut and palm kernel) have been discussed in the previous sections. Other commodity oils available at lower levels are cottonseed, ground-nut, and corn oil. The fatty acid composition of these and some other vegetable oils available in smaller amounts are given in Table 2.6. Many other minor oils are also available but information about these must be sought on appropriate websites or in sources pro-vided in the Reference list.

Almond, hazelnut, macadamia, and moringa oils are rich in oleic acid. Indeed hazelnut oil is so similar to olive oil that it is used to adulterate olive oil. Macadamia oil is rich in palmitoleic acid (16:1) as well as in oleic acid so that its content of monoene acids is

Table 2.6 Typical fatty acid composition of a range of vegetable oils available at lower levels

	16:0	18:1	18:2	18:3	Other
Almond		65–70			
Avocado	10–20	60–70	10–15		
Camelina		10–20	16–24	30–40	
Corn	11	25	60		
Cottonseed	23	17	56		
Groundnut	11	53	27		C_{20-24} 6
Hazelnut		74–80	6–8		
Hemp	4–9	8–15	53–60	15–25	
Linseed	6	19	24	47	
Linola	6	16	72	2	
Macadamia		55–65			16:1 16–23
Moringa	12	72	2		C_{20-26} 11
Perilla		13–15	14–18	57–64	
Rice bran	20	42	32		
Safflower		14	75		
HO saff		74	16		

Note: 16:0, 18:2, and other acids may also be present where no values are cited.

very high. Safflower seed oil is normally a linoleic acid-rich oil but a high-oleic variety is also available. Linseed oil (and perilla oil) is a convenient rich source of linolenic acid. Most linseed oil is used industrially but some high-grade oil finds food uses as a convenient source of omega-3 acid, albeit at the C_{18} chain length. There is also some food consumption of this material as seeds. When used for food purposes the name linseed is replaced by flax. A mutant of linseed resulting from chemical mutation produces a useful high-linoleic acid oil called linola and used as an alternative to sunflower oil. Camelina oil contains useful levels of linoleic and linolenic acids and is incorporated in spreads as a source of omega-3 acid. Rice bran oil is a significant oil, particularly in rice producing countries. It contains comparable levels of oleic and linoleic acids and has its own powerful antioxidants.

2.11   Single cell oils

Oils and fats produced by the agricultural supply industry come mainly from plant sources and also to a minor extent from animals. An alternative approach is to seek new lipid sources from microorganisms. Some of these can be made to produce high levels of lipids with an interesting fatty acid composition. While it is unlikely that these will replace the more conventional commodity oils for traditional use, nevertheless, they are already providing supplies of high-quality long-chain polyunsaturated fatty acids for infant formula and other special purposes. They are useful sources of arachidonic (20:4) and docosahexaenoic acid (22:6). The major source of DHA is fish oil but supplies of these are limited and some are

Table 2.7 Fatty acid composition of two commercial single cell oils

	14:0	16:0	16:1	18:0	18:1	18:2 (ω-6)	20:3 (ω-6)	20:4 (ω-6)	22:6 (ω-3)
ARASCO[a]	0.4	8	0	11	14	7	4	49	–
DHASCO[b]	20	18	2	0.4	15	0.6	–	–	39

Notes
[a]From *Mortierella alpine*.
[b]From *Crypthecodinium cohnii*.

contaminated with undesirable environmental pollutants. The fatty acid composition of two commercial products is detailed in Table 2.7. Serious attempts are being made to produce algal oils (of different fatty acid composition) that can be used as biodiesel. One of their attractions is the very high yield per hectare compared with more conventional agricultural products.

CHAPTER 2

Extraction, Refining, and Modification Processes

3.1 Extraction

This chapter outlines the steps whereby seeds, oil-containing fruits, or animal organs or carcases are changed from their harvested or available form to fat products that the food processor can incorporate into food. These include:

- extraction to isolate crude oil or fat;
- refining of the crude product, usually through several stages;
- modification of the native oil or fat to make it more suitable for its end-use.

During all the stages of extraction, refining, and modification, and also during transport and storage the oil must be protected against deterioration. Since this is most likely to be the consequence of hydrolytic or oxidative change requiring water and oxygen (air), respectively, these materials should be excluded as far as possible. Such changes are temperature dependent and oxidation can be promoted by light and by some metals. Oils and fats must therefore always be handled under appropriate conditions.

Extraction of oily fruits (palm, olive) involves pressing while extraction of seeds (kernels, beans) is achieved by pressing and/or solvent extraction with hexane or methylpentane – commonly called isohexane. After extraction of oil from seeds the residue is generally a protein-rich meal used as animal feed or as human food. This second component is an important part of the economics of oilseeds. Oil obtained by pressing is considered by some to be more natural and superior to solvent-extracted oil but most commodity vegetable oils, other than palm and olive oil, are solvent extracted. Animal fats are usually 'rendered' by heating with dry heat or steam. This is an important step in the economic, safe, and environmentally acceptable disposal of material remaining after the removal of butcher meat.

3.2   Refining

Crude oil is mainly a mixture of triacylglycerols along with the minor components detailed in Chapter 1, undesirable pigments, oxidation products, and metals. The purpose of the refining processes is to remove unwanted impurities while maintaining as far as possible the level of desirable minor components. Where these latter are removed it is often possible to trap them in a side stream and to recover them for alternative use. If the facilities for doing this are not to hand then the minor components may be added back to the meal, incorporated into animal feed, or used as fertiliser.

In Europe the term 'refining' is applied to the whole series of processes described here but in the USA it tends to be equated with removal of free acid (neutralisation).

- Degumming involves treatment with water or dilute acid (phosphoric or citric) producing a gum containing phospholipids (and trace metals) which can be separated with a centrifuge. The phospholipids are recovered as crude lecithin. Degumming is usually linked with extraction rather than with the subsequent refining processes. Phospholipids not easily removed by these degumming procedures are said to be 'non-hydratable' (NHP) and are mainly phosphatidic acids and lysophosphatidic acids present as calcium or magnesium salts. Other degumming procedures involve the use of ethylene diamine tetra-acetic acid for removal of trace metals or the use of appropriate enzymes. In the latter, appropriate lipases promote splitting of the phospholipids into forms that are more readily separated and removed.
- Neutralisation frequently requires treatment with aqueous alkali to remove free acids. Some oil is lost along with the soaps in a (large) water stream that has to be disposed of in an environmentally appropriate manner. Free acids are also removed during deodorisation by steam distillation (physical refining). This latter method is increasingly favoured because of its reduced environmental demand.

- Bleaching requires heating the oil at 80–180°C (but mainly at 90–120°C) with acid-activated bleaching earth. As the name implies the process was designed to remove colour (carotenes, chlorophyll) but trace metals are also removed along with soaps and residual phospholipids. However, under the acidic conditions free sterols may be dehydrated to steradienes and *cis* bonds in fatty acids may change to the *trans* form. Bleaching is the most expensive refining step because of the cost of obtaining the bleaching earth, of disposing of spent earth, and through loss of some oil. Polycyclic aromatic hydrocarbons (PAH) when present in the oil are not removed by bleaching. Volatile members are removed during deodorisation but the non-volatile PAH are removed by adsorption on activated carbon added to the bleaching earth.

- Deodorisation is the final refining step and requires the oil to be sparged with steam at 170–250°C under reduced pressure to remove oxidation products responsible for off-flavour. At higher temperatures (>220°C) there is isomerisation of *cis* to *trans* bonds so highly unsaturated oils should be deodorised at the lowest possible temperature. This applies especially to fish oils with their highly unsaturated long-chain PUFA which should not be heated above 180°C and to soybean and rapeseed oils which contain linolenic acid. Polyunsaturated fatty acids only retain their important nutritional properties if they remain in the all-*cis* forms. Deodoriser distillate is itself a valuable by-product that serves as a source of sterols and tocopherols.

The sequential processes of degumming, neutralisation, bleaching, and deodorisation are called chemical refining in distinction to the alternative of physical refining which requires only bleaching and deodorisation. The latter is a useful alternative mainly for oils (like palm oil) with low levels of phospholipids. The procedure is environmentally attractive because it avoids the large volume of waste water associated with neutralisation and financially attractive because there is less loss of oil. It is applied particularly to palm oil but is being used increasingly for other oils combined with improved degumming procedures.

The products of these processes are described as RBD (refined, bleached, and deodorised) oils. Specifications are defined by the Federation of Oils, Seeds and Fats Associations Ltd (FOSFA International) in Europe, by the USA-based National Institute of Oilseed

Products (NIOP), by the Palm Oil Refiners' Association of Malaysia (PORAM), and by the American Soybean Association (ASA). These internationally recognised standards cover a range of physical and chemical characteristics including free fatty acid content, iodine value, moisture and impurities, and colour.

3.3 Modification processes

The food processor has access to only a limited number of refined commodity oils and none of these may be ideal for purpose. As a consequence, procedures have been developed to modify the oils. Before describing these processes it is useful to consider in what ways the natural oils may be inadequate.

The ideal oil or fat should have physical, chemical, and nutritional properties appropriate for its end-use. However, these requirements are not always mutually compatible and compromises must be made. For example, the desired physical and chemical properties may only be achieved with some loss of nutritional quality. Typically, physical properties related to melting behaviour and crystalline form are important in a spread and to salad oils which should be free of solid. The most important chemical property is oxidative instability leading to the lower shelf life and resulting particularly from the presence of linolenic acid in unmodified soybean and rapeseed oils. Nutritional properties relate to the levels of saturated acids, to acids with *trans* unsaturation, and to the level and nature of the polyunsaturated fatty acids. These properties are detailed in Chapter 7.

Methods of modification may be technological or biological. Technological procedures include blending, fractionation, hydrogenation, and interesterification with chemical or enzymatic catalysts. Biological procedures include the agricultural development of new (minor) crops to make them more suitable for commercial growing and harvesting, seed breeding by conventional or newer procedures to produce oils with a more desirable fatty acid composition, and production of single cell oils rich in valuable polyunsaturated fatty acids not otherwise easily available.

In the four technological procedures listed above consideration has to be given to their effectiveness in achieving the desired physical, chemical, and nutritional properties and to their relative costs. Blending is the cheapest and hydrogenation is the most costly in

hydrogenation (known as brush hydrogenation) is applied to oils containing linolenic acid (soybean oil and rapeseed oil) to reduce the level of this triene acid to about one half its normal value thereby extending the shelf life of foods containing these oils. At the other extreme, oils are virtually completely hydrogenated to iodine values below 2. Of greater importance is partial hydrogenation applied to soybean and other linoleic-rich oils to raise the content of solid triacylglycerols. This is achieved through formation of saturated acids and of *trans* unsaturated acids.

Partial hydrogenation of an unsaturated oil gives a product of higher melting point (more suitable for spreads and cooking fats) and of enhanced oxidative stability through having less polyunsaturated fatty acid. These benefits are only achieved at some nutritional cost. The level of essential fatty acid (PUFA) is lowered and acids with *trans* configuration are produced. These modifications follow the molecular changes resulting from partial hydrogenation including saturation of some unsaturated centres, stereomutation of unsaturated centres (conversion of *cis* to *trans* isomers), double bond migration, and conversion of linoleate mainly to *trans* 18:1 isomers.

$$18:3 \rightarrow 18:2 \rightarrow 18:1 \ (cis \text{ and } trans \text{ isomers}) \rightarrow 18:0$$

3.7 Interesterification using a chemical catalyst

Interesterification is a procedure for rearranging the fatty acids in an oil or in a blend of oils so that triacylglycerol composition is changed. The fatty acid composition of the single oil or the blend remains unchanged. Usually the blend will contain two or more oils differing in chain length and/or in patterns of unsaturation. With an alkaline catalyst, such as NaOH or NaOMe, fatty acids are randomly distributed in the product in contrast to the natural vegetable oils that are produced by enzymatically catalysed biological processes and where the fatty acids are not randomly distributed. These changes in triacylglycerol composition affect thermal behaviour and may also affect bio-availability by reason of what fatty acids are in the *sn*-2 position. There is some loss of oil through conversion to acid (with NaOH) or ester (with NaOMe).

This procedure is being used to produce fats suitable for use as spreads without hydrogenation. A soft (polyunsaturated) oil is blended with hardstock and the mixture is interesterified. The hardstock is either a suitable palm stearin or a fully hydrogenated oil. Unfortunately with the latter mixture it will probably require the term 'hydrogenated' on the label. Even though the fully hydrogenated oil will contain virtually no *trans* acids the process itself is now considered undesirable – mainly on the basis of perception rather than science.

3.8   Interesterification using an enzymatic catalyst

Interesterification can also be carried out with lipases. The enzymatic processes have advantages over the chemical process in that they occur under milder conditions, may require less costly equipment, and produce less by-product so that there is less waste and less effort is required to purify the product. However, the major benefit of using a lipase is the added control over the nature of the product as a consequence of the specificity shown by many lipases. Lipases with specificity relating to fatty acid chain length or double bond position in the acyl chain can be used to confine changes to a particular group of acids while other lipases are specific for glycerol esters (mono-, di-, or tri-acylglycerols) or distinguish between fatty acids attached to the differing the different hydroxyl groups in glycerol. Many lipases are described as being 1,3-specific implying that changes can be made at glycerol positions 1 and 3 but not at position 2 where the ester group remains unchanged. Though lipase preparations are becoming more stable and less expensive the cost of enzyme-catalysed processes is still a major disadvantage. However, the enzymatic reactions are perceived as the more natural and therefore attractive to the 'green' lobby. Compounds attracting a lot of interest are of the type MLM where M represents a short-chain acid (frequently C_8) which is easily metabolised and where L represents a long-chain acid such as eicosapentaenoic acid (EPA) or docosahexaenoic acid (DHA). These important acids attached to the 2-position are thereby made bio-available.

With either chemical or enzymatic catalysts interesterified products are generally less stable than the original oils probably through changes – not yet fully understood – in the balance of pro- and anti-oxidants. There is a perception in some quarters that the enzymatic reaction is superior because it is 'more natural' and avoids the use of 'chemicals'.

3.9   Domestication of wild crops

The oil and fat business is based almost entirely on a limited number of commodity oils differing in fatty acid composition but there are many other plant species with fatty acid composition not very different from the commodity oils. These could be used as food lipids but there would have to be a special reason for developing them through the long chain of events from agronomical improvements to retail marketing. A few of these minor oils were discussed in Chapter 2. There are also plants producing uncommon acids such as epoxy acids, acids with conjugated unsaturation, or oils with a very high level (>80%) of a single acid. Attempts to domesticate and commercialise such plants and their seed oils take a long time to develop with some niche products of this kind taking 20 and more years to bring to market. Most, but not all, of these oils are of interest to the oleochemical rather than the food industry.

3.10   Oilseeds modified by conventional seed breeding or by genetic engineering

Because of the difficulties in domesticating wild plants greater effort has been directed to the modifying of plants that are already grown and harvested on a commercial scale and where good agronomic procedures are already well developed. This has the disadvantage of minimising the range of important plant species thereby limiting biodiversity. The changes to be sought are partly agronomic such as reduced use of herbicide and pesticide but they include changes in fatty acid composition, triacylglycerol composition, and in levels of minor components. These changes must be achieved without sacrifice of yield and must be biologically stable from season to season. They have to be accompanied by procedures of identity preservation. The modified seed must be kept separate at all times from its more conventional form. This has consequences for harvesting, transporting, and extracting the seed and for the subsequent handling of the oil. Such changes may be brought about by conventional seed breeding or by newer procedures of genetic engineering. It is important to know which method has been used because of the concerns expressed by some communities about procedures involving transgenic modification. This objection is adding to costs and is becoming harder to justify being based more on perception than on science. Oilseeds from a known source and kept distinct from seeds produced and handled in a different way are said to be 'identity preserved' (IP).

Changes of fatty acid composition which have been sought include: reduced levels of saturated acids for nutritional reasons, reduced levels of linolenic acid and/or higher levels of saturated acids to avoid hydrogenation (with consequent production of undesirable *trans* acids), and higher levels of oleic acid (see Table 2.5). One important and exciting possibility is to develop plant systems that will produce long-chain polyunsaturated fatty acids such as arachidonic acid (20:4), EPA (20:5), and DHA (22:6). There have been interesting developments in a number of research laboratories but such plants will probably be genetically modified and are 10–20 years from commercial development.

In view of the high demand for oils and fats at this time for food and non-food purposes it is important to increase supplies. This may be achieved by seed breeding to give higher yields, by developing seeds, which are more drought-resistant and will so give better yields under adverse conditions, and seeds, which will grow under harsher conditions of climate (lower temperatures or shorter growing season) or of soil (high salinity). The last could also be irrigated with poorer quality water.

3.11 Animal fats modified through nutritional changes

From a nutritional viewpoint land animal depot fats are perceived as having several disadvantages. They are generally rich in serum cholesterol-raising saturated acids such as myristic and palmitic, they often contain acids with *trans* unsaturation, and they have high levels of cholesterol. Also their level of essential fatty acids is low and they contain little if any antioxidant. Further, animal fats are not acceptable to vegetarians and to some ethnic groups. Nevertheless animal fats contain low levels of long-chain PUFA and are a valuable source of such acids in the human diet especially for those who consume little or no fish.

Some of the perceived disadvantages in ruminant animals are the consequence of biohydrogenation processes taking place within the rumen and dietary regimes have been proposed to circumvent these changes. There is also an interest in modifying the fatty acid composition of chicken eggs and meat by appropriate changes to the diet of the chicken. This has been seen as a way of enhancing the (human) dietary intake of conjugated linoleic acid (CLA) and of EPA and DHA (long-chain omega-3 acids). Care has to be taken that these changes in fatty acid composition do not reduce the shelf life of the products and that they do not lead to unexpected flavours and odours when cooked.

Analytical Parameters

4.1 Introduction

Those working in the food industry need to have some knowledge of lipid analysis. Some properties will be detailed in a specification, others relating to starting materials or food products may have to be measured in-house or in an external laboratory. In whatever form the information comes and from whatever source it comes the food scientist needs to know what it signifies and whether the information provided is acceptable or not. Standardised and widely-accepted analytical procedures are preferred to methods that are not generally practised in other laboratories. If goods are to be traded internationally analytical procedures must be robust and widely recognised. What follows is not a detailed account of these procedures but rather an outline of methods and the value of results so obtained. Other texts concentrate on this topic and the fullest account at the present time is Christie's recent book (2003) and his website *The Lipid Library*.

Traditional procedures of analysis were essentially chemical in nature. They involved chemical reagents and solvents, they were generally labour-intensive, and often required gram quantities of material. Some of these still have a place but increasingly they have been replaced by procedures based on physics rather than chemistry, particularly chromatography and spectroscopy. These latter are generally quicker, less labour-intensive, more accurate, and require less material. However, equipment is more sophisticated and more expensive. Spectrometers and chromatography systems have largely replaced burettes and the older type of glass pipettes.

Organisations such as those listed below provide similar but not identical directions.

AOAC   The Association of Official Analytical Chemists
AOCS   The American Oil Chemists' Society
BSI   The British Standards Institution
ISO   The International Organization for Standardization
IUPAC   The International Union of Pure and Applied Chemists
          European Pharmacopoeia, 5th edition, Council of Europe, Strasbourg, 2004

Before any test is carried out it is necessary to obtain a representative sample of material and perhaps to transport and store this before any measurement is made. There are standard procedures for all these stages. Attention must be given to the temperature of storage, the nature of the container, the inhibition of enzyme activity, and the possible addition of antioxidants. Unless care is taken in all these matters even the most careful analysis will be valueless.

4.2   Oil content

Different ways of quantitatively extracting lipid from a sample are available and depend on the nature of the matrix in which the lipid exists.

For oilseeds, the oil is generally extracted from crushed seed by the Soxhlet procedure using hexane or other suitable hydrocarbon fraction such as that boiling between 40°C and 60°C. This method provides a sample of oil that can also be used for further tests. Non-destructive methods suitable for routine assessment of many samples are based on near infrared (NIR, Section 4.8) or nuclear magnetic resonance (NMR, Section 4.9). Extraction of oils and fats on an industrial scale is described in Section 3.1.

More complex methods are required for biological sources such as liver or blood, often associated with a high proportion of water. In foodstuffs lipid is accompanied by protein and/or carbohydrate and such sources may also require special procedures (McLean and Drake, 2002).

Biological samples are extracted with chloroform–methanol according to the well-established methods of Folch *et al.* (1957) or of Bligh and Dyer (1959) as described by Christie (2003). In the Folch extraction, ground or homogenised tissue is shaken with a 2:1 mixture of chloroform and methanol and the organic extract is subsequently partitioned with aqueous potassium chloride solution. The combined layers should have a volume ratio of 8:4:3 (chloroform/methanol/water). Extracted lipid is in the (lower) chloroform layer. The Bligh and Dyer method was developed for fish muscle and other wet tissue assumed to contain about 88 g of water in every 100 g of tissue. The tissue (100 g) is homogenised with chloroform (100 ml) and methanol (200 ml) and, after filtering, residual tissue is homogenised a second time with chloroform (100 ml). The two organic extracts are combined and shaken with aqueous potassium chloride (0.88%, 100 ml). After settling, the lipid partitions into the lower layer.

Fat in food has been defined in Europe as total lipids including phospholipids and in the United States as fatty acids from monoacylglycerols, diacylglycerols, triacylglycerols, free acids, phospholipids, and sterol esters. These assessments have generally been made by extraction with an appropriate solvent assuming that all lipid is extracted and that the extract is only lipid. There may be problems when lipid is associated with protein or with carbohydrate and modified methods are needed. An alternative method is to hydrolyse the total sample with acid or alkali after adding triundecanoin (glycerol ester of undecanoic acid – 11:0) as internal standard. The resulting fatty acids are extracted, converted to methyl esters, and examined by gas chromatography (GC). The results are converted to triacylglycerol equivalents and expressed as fat (McLean and Drake, 2002).

4.3   Unsaturation – iodine value

Oils and fats contain saturated and unsaturated acids and many of their properties depend on the ratio of these two types of acids. Traditionally, average unsaturation has been measured as the iodine value based on chemical reaction with iodine monochloride (Wijs' reagent) or other mixed halogen compound under controlled conditions. The value is still cited in many specifications relating to oils and fats. However, it has a number of disadvantages and limitations. The measurement is time-consuming, labour-intensive, and uses undesirable reagents and solvents. For this reason iodine value is now often calculated from the fatty acid composition determined by GC using the theoretical iodine values of individual components.

The calculated iodine values of methyl stearate, oleate, linoleate, and linolenate are 0, 85.6, 173.2, and 260.3 based on the function:

$$25380 \times \text{(number of double bonds)} \div \text{molecular weight}$$

However, the agreement between observed and calculated values is not good because (a) no allowance is made for unsaponifiable material which generally contains olefinic compounds and (b) the GC trace may contain minor peaks which are unidentified and ignored. Also, measured iodine values of polyunsaturated fatty acids may be low through incomplete halogenation. An important limitation is that the iodine value does not distinguish between *cis* and *trans* isomers and this information is important when following partial catalytic hydrogenation.

Knothe (2002) has drawn attention to the fact that average unsaturation distinguishes between saturated and unsaturated acids but does not reflect the important difference in reactivity between monounsaturated and polyunsaturated acids. He has suggested new indices measuring the allylic position equivalent (APE) from monounsaturated and polyunsaturated acids and the bis-allylic position equivalent (BAPE) from polyunsaturated acids only. These can be determined by GC or from ^{1}H and ^{13}C NMR signals characteristic of each of these acid types (Sections 4.9 and 4.10). This suggestion has not been widely applied but the principle remains valid (see the concept of oxidisability in Section 6.2).

4.4   Saponification – free acids, sap value

The level of free acid is listed in most specifications for crude and refined oils. It is measured by titration with standard sodium hydroxide solution and may be expressed as acid value (mg KOH required to neutralise 1 g of fat) or as percentage of free fatty acid. For most oils free fatty acid is expressed as oleic acid and is equal to the acid value divided by 1.99 (usually rounded to 2.0). Free acids present in crude oils are largely removed during the refining processes and the acid value should be below 0.1%.

The amount of alkali required to hydrolyse (saponify) a fat is a measure of the average chain length of the acyl chains though this value is affected by unsaponifiable material also present in the oil.

This parameter may be reported as 'saponification value' (SV) or 'saponification equivalent' (SE). These numbers are inversely related by the expression SE = 56,100/SV. The SE is the average molecular weight of all the acyl chains. With increasing chain length, SE rises but SV falls. Typical SVs for some common oils include coconut 248–265, palm kernel oil 230–254, cocoa butter 192–200, palm oil 190–209, cottonseed 189–198, soybean 189–195, sunflower 188–194, corn 187–195, groundnut 187–196, olive 184–196, and rape 182–193. High values are associated with the two lauric oils and the lower values with oils rich in C_{16} and C_{18} acids.

When a natural fat or oil is hydrolysed it gives fatty acids (soluble in aqueous alkali), glycerol (soluble in water), and other material (insoluble in aqueous alkali). This last can be extracted with an appropriate organic solvent (hexane or diethyl ether) and is described as unsaponifiable or non-saponifiable material. It includes sterols, tocopherols, hydrocarbons, long-chain alcohols, etc. There is a growing interest in these compounds and chromatographic and/or spectroscopic methods of analysing this fraction in more detail are available. Unsaponifiable material is normally less than 2% of the total oil though sometimes it will be higher. Wax esters are hydrolysed to long-chain acids and alcohols and the latter will be part of the unsaponifiable fraction. Spreads with added phytosterol esters (Sections 7.7 and 8.3) will also have elevated levels of unsaponifiable material.

4.5   Melting behaviour, solid fat content, low-temperature properties

Fats are not pure organic compounds with sharp melting points but mixtures of many individual triacylglycerols each of which may be solid or liquid at ambient temperature. Most spreads are plastic solids that deform under pressure as during the spreading operation because they are mixtures of solid and liquid components. The proportion of these two phases varies with temperature and it is frequently necessary to know the solid/liquid ratio at a range of selected temperatures. This is important in assessing the quality of spreading fats and confectionary fats (Sections 8.3 and 8.7). The temperature at which solid first appears on cooling is also important in frying oils and in salad oils (Sections 8.5 and 8.6). The 'titre' denotes the solidification point (°C) of the fatty acids derived from a fat while the slip melting point is the temperature at which a column of fat ($10 \pm 2\,mm$), contained in an open capillary tube and immersed in water to a depth of 30 mm, starts to rise. This is a useful low-temperature property (see also Sections 5.1–5.3).

Of greater value is the measurement of solid fat content by low-resolution (pulse) ^{1}H NMR spectroscopy (Section 4.9). The percentage of solid determined by pulse-NMR is based on the ratio of the response from the hydrogen nuclei in the solid phase and that from all the hydrogen nuclei in the sample. Measurements made at a range of temperatures give a plot of solid content against temperature. The slope of this line and the temperature at which there is no solid phase provide useful information about the melting and rheological behaviour of the sample under investigation.

4.6   Oxidation – peroxide value, anisidine value, stability, shelf life, stability trials, taste panels

In common with other olefinic compounds oils and fats react with oxygen. The process is complex (Section 6.2) and usually undesirable. Two major questions are asked of the analyst in this connection: how far has the sample already been oxidised and how long will the (food) sample last before it is unacceptable? (*i.e.* What is its shelf life?) The first requires a measurement of present status while the latter requires a predictive measurement. The most common oxidative process is autoxidation. This occurs with an induction period during which deterioration is not severe and it is useful for food producers to know the length of this. Several stages of oxidation are recognised and tests are available for each stage:

- Primary products of oxidation are allylic hydroperoxides and are measured as peroxide value or as conjugated dienes formed during oxidation of PUFA.
- Secondary products are mainly unsaturated aldehydes and are measured by the anisidine value.
- Tertiary oxidation products include short-chain acids measured by the Rancimat or oil stability index (OSI) or malondialdehyde measured by the TBA test.

Though knowledge of oxidative deterioration is most important for goods stored at ambient or refrigerator temperatures the changes can be accelerated at elevated temperatures. Unfortunately reaction at higher temperature is not always a good predictor of reactions occurring at lower temperatures.

The most common method of assessing oxidative status is by measurement of hydroperoxides. These molecules react with acidified potassium iodide to liberate iodine that can be measured volumetrically by reaction with sodium thiosulphate. The value represents mmol of oxygen per 2 kg of fat and this means that ~0.1% of the olefinic molecules in an oil have been oxidised when the peroxide value is 2. Freshly refined material should have a peroxide value below 1. A fat is considered to be rancid at a peroxide value exceeding 10. Refining destroys hydroperoxides but it does not regenerate the fat in its original form. Hydroperoxides are cleaved to aldehydes during refining. While volatile aldehydes are removed during subsequent refining short-chain aldehydes attached to the glycerol moiety remain (sometimes called core aldehydes) and can be detected by the anisidine value. Refining an oil that has already been oxidised will therefore reduce the peroxide value but the anisidine value will not be reduced to zero. These two measurements may be combined in a totox value representing the sum of twice the peroxide value plus the anisidine value.

The anisidine value is based on the measurement of the intensity of the chromophore at 350 nm arising from molecules of the type $ArN{=}CHCH{=}CHR'$ produced by reaction of anisidine (4-$MeOC_6H_4NH_2$ represented as $ArNH_2$) with carbonyl compounds which are mainly 2-enals ($R'CH{=}CHCHO$). This value varies depending on the enals actually present and is therefore only strictly comparable across results for a single type of oil. An anisidine value of 1 corresponds with ~0.1% of oxidised material.

$$ArNH_2 + OCHCH{=}CHR' \rightarrow ArN{=}CHCH{=}CHR'$$

Early stages of autoxidation can also be detected by measurement of ultraviolet absorption at 234 nm resulting from conjugated dienes formed during oxidation of polyunsaturated fatty acids. This method is not suitable for heated fat, for fat that already contains conjugated dienoic acids, nor for fats with a high content of oleic acid and consequent low levels of linoleic acid (Table 6.2).

$$CH_3(CH_2)_3 CH_2 CH = CH CH_2 CH = CH CH_2 (CH_2)_6 COOH$$

Figure 4.1 Linoleic acid with two allylic groups (italic) and one bis-allylic group (bold).

In the Rancimat and Omnium Oxidative Stability measurements a stream of air is drawn through oil heated at 100–140°C into a vessel containing de-ionised water. Short-chain acids – mainly formic (HCOOH) – increase the conductivity of the water and the induction period is indicated by the time that elapses before there is a rapid increase in conductivity. These measurements may be of limited value for predicting the stability of a range of oils but for repeated samples of the same oil they give useful comparative values. They have largely replaced older active oxygen methods (AOM).

In the older accelerated tests (Schaal, Active Oxygen) the oil or fat was held at a temperature up to 100°C and the time taken to reach an arbitrary peroxide value was measured. This was taken as an indication of the induction period and hence shelf life under normal storage conditions.

In biological experiments the presence of short-chain hydrocarbons in breath may be measured. Ethane (C_2H_6) comes from omega-3 acids and pentane (C_5H_{12}) from omega-6 acids.

Headspace analysis may be carried out in various ways using GC to separate and identify short-chain compounds – mainly aldehydes – formed by decomposition of hydroperoxides. Compounds such as 4-heptenal, and the 2,6- and 3,6-nonadienals are considered to be the most significant flavour notes but many of the volatile materials have little sensory effect.

The ultimate assessment of food flavour and texture is achieved by taste panels. These are discussed by Malcolmson in Shahidi (2005).

4.7 Gas chromatography

Lipid analysis is now dominated by chromatographic and spectroscopic procedures based on physics rather than chemistry. Procedures are increasingly automated with results available in electronic form. Some companies outsource their analyses of oils and fats to specialist laboratories that have dedicated instruments and wide expertise.

Some chromatographic procedures (thin layer chromatography TLC, high-performance liquid chromatography HPLC, silver ion chromatography) and the mass spectrometric (MS) techniques that may be associated with them are more likely to be found in the research laboratory than in the quality control laboratory.

The most widely used property of any fat is its fatty acid composition. This indicates what fatty acids are present and at what level and is universally determined by GC of the methyl esters derived from the triacylglycerols. Minor components (sterols, sterol esters, and tocopherols) can also be determined by appropriate GC procedures. In considering fatty acid composition attention has to be given to the procedures for preparing the methyl esters, to the GC conditions, and to the ways in which the results are presented.

Triacylglycerols are easily converted to methyl esters by base-catalysed transesterification. This involves reaction with excess of methanol containing sodium methoxide and is complete in a few minutes at 50°C. It may be necessary to use a co-solvent such as toluene and antioxidant is generally added to protect the unsaturated esters throughout the analysis. This method works well with refined oils of low acidity but free acid is not esterified under alkaline conditions and when present, as in crude extracted oils, methyl esters are more generally made by acid-catalysed esterification and transesterification using methanol and hydrogen chloride, sulfuric acid, or boron trifluoride along with co-solvent and antioxidant. These may be combined in a process involving alkaline hydrolysis followed by acid-catalysed esterification. Special methods may be necessary when the oil contains fatty acids with acid-labile functional groups such as epoxide or cyclopropene.

GC is employed to separate and quantify component acids in the form of their methyl esters. This efficient separation procedure is based on partition chromatography in which the stationary phase is usually coated on the inner wall of a fused silica capillary tube 10–100m in length and is liquid at the temperature of analysis. This phase may be non-polar, weakly polar, or highly polar. The gas phase is usually nitrogen or helium or hydrogen in order of increasing resolving power. The column is heated to a range of temperatures limited only by the thermal stability of the stationary phase and of the analyte. Elution is slower at lower temperatures but separation is improved and it may be necessary to make a compromise between time of elution and efficiency of separation. Rapid procedures for GC separation have been described and are required when very large numbers of samples have to be examined. Separation is monitored with a flame ionisation detector that is remarkable for its robust nature and for its linear response over a wide range of concentration. The separation may be carried out at constant temperature (isothermal) or according to a prearranged programme during which the temperature is gradually raised. With automatic injection equipment can be organised to operate overnight without manual intervention.

The column allows partitioning of the separate constituents of the analyte between the stationary phase as a thin film on the inner surface of the capillary column and the mobile phase (gas). The individual components of the analyte travel down the column and are eluted after different times (retention time) depending on the proportion of time spent in the stationary and mobile phases. The efficiency of a chromatographic system depends on the nature and flow rate of the carrier gas, column dimensions, liquid phase thickness, and column temperature. These parameters are optimised within practical constraints such as the time that can be given to each analysis.

The peaks in the chromatogram are identified on the basis of their elution time in comparison with data obtained from standard mixtures of esters under identical chromatographic conditions. If there is any doubt about the identification it may be necessary to repeat the GC with a column of differing polarity or to combine the GC separation with MS for structural identification.

The simplest way of reporting the results is to normalise all peaks (*i.e.* to express the area of each peak as a percentage of the total area under all peaks). Results should then total 100.0 even if some of the smaller peaks have not been identified with certainty. The GC may fail to detect volatile impurities (present in the uncounted solvent peak) and components which are not eluted during the time given to the analysis. These include oxidised and polymerised impurities and some minor components in the oil. Increasingly therefore results are expressed as milligram/gram. These can be determined with the help of an internal standard – usually the triacylglycerol of an odd-chain acid with 11, 17, 19, or 23 carbon atoms – which has to be added in a weighed amount to the fat being analysed. The internal standard is then subject to the same chemical reactions and extraction procedure as the sample being examined. The results should approach a total of 1000 but are more likely to be around 900.

4.8   Near-infrared and Fourier transform infrared spectroscopy

The near-infrared region of the spectrum, composed of overtones and combinations of the fundamental bands, was considered unimportant until developments in computing made it possible to exploit this information. Near-infrared reflectance spectroscopy (NIRS), based on commercial instruments, is now much used in agriculture and beyond. It is used, for example, to determine the content of moisture, protein, and fat in a batch of seeds. Its use has been extended to the determination of fatty acid composition and this may be carried out on a single seed. This is of great benefit in breeding programmes. The procedure is rapid, non-destructive, and involves neither sample grinding nor chemical modification. Calibration equations based on a large number of samples are required but instruments from different laboratories can be integrated in a network with calibration equations developed on a master instrument and then used in all the satellite instruments in the network.

Fourier transform infrared (FTIR) spectroscopy has advanced dramatically in recent years and is now used as an alternative way of measuring several properties important for lipid analysts. An FTIR spectrometer can record the entire infrared spectrum in one second and this can be added to many other scans through a fast Fourier transform algorithm to produce a conventional infrared absorption spectrum. Spectra based on interferometry have several advantages over spectra from more conventional dispersive instruments. There is a marked improvement in signal to noise ratio, higher energy throughput, superior resolution, and greater wavelength accuracy through the use of an internal laser. Undiluted edible oils are particularly suited to FTIR analysis as they are liquid, easy to handle, and have relatively simple spectra. Preliminary calibration is necessary to convert spectral information into useful data and once this is available the system may be used to measure parameters such as *cis–trans* ratios, iodine value, saponification number, free acid content, peroxide value, and anisidine value. Details are available on the website and in the paper by Tseng and Wang (2007).

With developments in MS procedures the mass spectrometer itself is used as a quantitative instrument. The charge for individual peaks is compared with total ion current (Section 4.10). The MS procedure has the added advantage of providing structural information about the material in the peak. This is in contrast to FID measurement where structure can only be inferred (though with a large measure of certainty in routine analyses).

4.9 ^{1}H NMR spectroscopy

^{1}H NMR spectroscopy is used in two ways in the study of lipids. With wide-line (low resolution or pulsed) instruments it is possible to determine the proportion of solid and liquid in a fat and the content of oil in a seed. High-resolution spectrometers, on the other hand, are used to examine solutions and give information about the solute, which may be an individual compound or a mixture, such as a natural oil or fat. Solids can also be examined when the spectrometer is used in 'magic angle' mode.

Low-resolution ^{1}H NMR or time-domain NMR is much used in quality control laboratories for the measurement of solid fat content, simultaneous determination of oil and moisture content, the study of oil and water droplet size distribution, and measurements that can be made through packaging. The technique has been reviewed by Meeussen in Hamilton and Cast (1999), Todt *et al.* in Dobson (2001), and Knothe (2003).

The ratio of solid and liquid phases in a fat is important in chocolate manufacture and in the understanding of melting behaviour in spreads (Sections 8.3 and 8.7). Low-resolution NMR has almost completely replaced the older method of dilatometry to measure solid fat content. The percentage of solids is given by the expression:

100 (hydrogen nuclei in the solid phase) ÷ (all the hydrogen nuclei in the sample)

These two types of hydrogen environment can be distinguished by observation of the relaxation signal. The signal for hydrogen atoms in solids decays quickly – less than 1% remains after 70 μs – while that from liquids decays very slowly requiring about 10,000 s. There are practical reasons why measurements cannot be made at the instant of the pulse and are usually made after 10 μs $(S_S + S_L)$

and after 70 μs (S_L only). Because some of the S_S signal will have already decayed after 10 μs the observed value has to be corrected by a factor determined by calibration of the system using samples of solid plastic (35–70%) in liquid paraffin.

These measurements require only about 6 s and are used routinely for the study of spreads and confectionery fats. However, fats needing polymorphic stabilisation such as cocoa butter have to be equilibrated before measurements are made and a tempering routine requiring up to about 40 h has been described.

By further adaptation the NMR system can be modified to distinguish between oil and moisture and it is possible to measure the oil and moisture content of around 1000 samples of seeds per day.

High-resolution spectroscopy, on the other hand, is used to examine solutions and gives information about the solute, which may be an individual compound or a mixture, such as a natural oil or fat.

A typical ¹H spectrum is shown in Figure 4.2. It contains signals that can be distinguished by chemical shift, coupling constant, splitting pattern, and area. The last of these provides quantitative information that can be presented as mol % in contrast to GC data given in wt %. The remaining parameters give structural information (Diehl in Dobson, 2001; Knothe, 2003; and *The Lipid Library*).

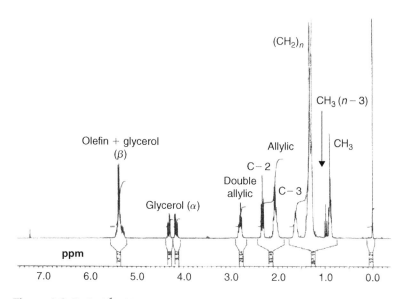

Figure 4.2 Typical ¹H NMR spectrum of a vegetable oil.

A saturated long-chain methyl ester has five signals with the following chemical shifts (ppm), number of hydrogen atoms, and splitting pattern:

•	CH_3	0.90	3	triplet
•	$(CH_2)_n$	1.31	2n	broad (many overlapping signals)
•	$-CH_2CH_2COOCH_3$	1.58	2	quintet
•	$-CH_2CH_2COOCH_3$	2.30	2	triplet
•	$-CH_2CH_2COOCH_3$	3.65	3	singlet

Such a spectrum indicates the presence of a straight-chain saturated methyl ester but does not distinguish between homologues in a mixture. In olefinic esters there are additional signals corresponding to the:

- olefinic hydrogen atoms ($-CH=CH-$ 5.35 ppm, 2H for oleate, 4H for linoleate, 6H for linolenate)
- allylic hydrogen atoms ($-CH_2CH=CHCH_2-$ 2.05 ppm, 4H)
- doubly allylic hydrogen atoms ($=CHCH_2CH=$ 2.77 ppm, 2H for linoleate, and 4H for linolenate)
- hydrogen atoms of the end methyl group of omega-3 esters produce a triplet at 0.98 ppm which is distinguished from the usual triplet at 0.90 ppm.

Glycerol esters have five hydrogen atoms associated with the glycerol unit. There is a one-proton signal at 5.25 ppm ($CHOCOR$) overlapping with olefinic signals and a four-proton signal split between 4.12 and 4.28 ppm (CH_2OCOR).

For vegetable oils containing the usual mixture of saturated acids and C_{18} unsaturated acids useful information can be obtained by ^{1}H NMR procedures that are non-destructive and require no chemical reactions. The signal at 2.30 ppm (α-methylene function) provides a measure of all the acyl groups. The signals at 0.90 and 0.98 ppm distinguish linolenate (omega-3) from all other esters. Signals at 2.77 ppm are a combined measure of triene (linolenate) and diene (linoleate) and those at 2.05 ppm relate to all of linolenate, linoleate, and oleate. The intensity of these signals can be used to calculate the composition (mol %) in terms of oleic, linoleic, linolenic, and total saturated acids but the results are less

accurate than those obtained by GC. The accuracy of this procedure is limited because the values are not determined independently
but are dependent on each other. For example, an error in measuring
omega-3 trienes will introduce errors in the subsequent assessment
of diene and monoene esters.

4.10 ^{13}C NMR and ^{31}P NMR spectroscopy

^{13}C NMR spectra are based on natural ^{13}C atoms present at a level
of 1.1% in organic compounds. The spectra provide two kinds of
information: the chemical shift of each signal (up to 50 signals in a
natural mixture of triacylglycerols) and their relative intensities. The
former is of qualitative value and permits identification of important
structural features. The latter, with appropriate safeguards, provides quantitative information of analytical value. Chemical shifts
may vary slightly with concentration of the solution under study and
(rather more) with the solvent employed. Most measurements are
made with solutions of about 1M concentration and CDCl$_3$ is the
solvent most commonly used. Other solvents include (CD$_3$)$_2$SO,
C$_6$D$_6$, and mixtures of CD$_3$OD and CDCl$_3$. Figure 4.3 shows a typical spectrum for a vegetable oil.

In using ^{13}C NMR data (chemical shifts and intensities) the first
step is to assign as many of the chemical shifts as possible. If
the substance under study is a mixture, many individual signals

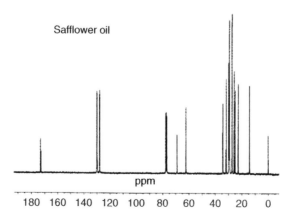

Safflower oil

ppm

180 160 140 120 100   80   60   40   20   0

Figure 4.3 ^{13}C NMR spectrum of safflower oil.

will appear as clusters. This makes interpretation more difficult but eventually provides additional information. It is wise to ignore signals in the methylene envelope (29.4–29.9 ppm) resulting from mid-chain carbon atoms that are not greatly influenced by nearby functional groups. Instead, examine the easily recognised shifts (in ppm) associated with the following carbon atoms $\omega 1$ (around 14.1), $\omega 2$ (22.8), $\omega 3$ (32.1), C-1 (174.1), C-2 (34.2), C-3 (25.1), glycerol (68.9 and 62.1), olefinic (127–132), and allylic (27.3 and 25.6).

The chemical shift of a carbon atom depends on its total environment to a distance of six or more atomic centres. For example, in glycerol trioleate the signals for the olefinic carbon atoms (C-9 and C-10) differ from one another and also, to a small extent, on whether the oleate is an α or β chain (attached to primary or secondary glycerol hydroxyl groups). The C-1 signal is also slightly different for saturated and $\Delta 9$ unsaturated chains. In these examples the difference is produced by structural changes up to 11 atomic centres away. This makes the spectrum more complex but also more informative when all the chemical shifts have been assigned.

In another example, the methyl groups at the end of the acyl chains in glycerol tripalmitate give one signal at about 14.1 ppm well separated from other signals and hence easily recognised. The difference between α and β chains for this signal in this molecule is too small to be observed but in a vegetable oil, containing saturated and unsaturated chains the peak at 14.1 ppm appears as a cluster of two or more signals. Each is indicative of a different environment for the methyl group and may result from omega-3, omega-6, or omega-9 acyl chains where the closest double bond affects the chemical shift of the methyl signal.

The signals for the acyl carbon atoms (C-1) in triacylglycerol mixtures appear as a complex cluster. One group of signals around 173.1 ppm and a second group around 172.7 ppm correspond to acyl carbon atoms in the (two) α-position and the carbon atom in the β-position, respectively, with peak areas in the ratio 2:1. These shifts do not differ greatly between saturated esters and those having unsaturation starting at the 9 position as in oleate and linoleate. However, different carbon shifts are observed when the double bond gets close to the acyl function as in $\Delta 4$ acids (DHA), $\Delta 5$ (EPA and AA), and $\Delta 6$ (petroselinic, γ-linolenic, and stearidonic). This makes it possible to determine the proportion of each of these acids in the α and β positions and this is now used in studies on fish oils which characteristically contain some of these acids (*European Pharmacopoeia*, 2005, 2006; Curtis in Breivik, 2007).

To obtain quantitative data attention has to be given to the protocol for obtaining the spectrum. In particular, the problem of relaxation has to be overcome either by adding a relaxation agent such as $Cr(acac)_3$ (chromium acetonylacetonoate) and/or by including a delay time between successive scans of the spectrum. This will add to the time required to collect the spectrum. Spectrometers now available operate at a frequency for ^{13}C of 68 MHz or more and spectra are generally obtained using an NOE-suppressed (nuclear Overhauser effect), inverse-gated, proton-decoupled technique. Exciting pulses have a 45–90° pulse angle and acquisition times (including delay times) are generally 1–20 s per scan. The number of scans is usually 1000 or more. The sample size for a routine ^{13}C NMR spectrum is normally 50–100 mg and the spectrum is obtained in 20–30 min. With smaller samples high-quality spectra can be obtained with as little as 1 mg but with a correspondingly longer acquisition time.

Assignments of chemical shift are often made on the basis of available knowledge. Existing information has been built up over the past 30 years assisted by the study of [2]H-containing compounds and the use of chemical shift reagents. Where the necessary information is not available more advanced spectroscopic procedures will assist assignment. This can also be made on the basis of line-width and relaxation measurements. Easily recognised carbon atoms present in most triacylglycerols have been cited above. This provides enough information to make a preliminary assignment to the signals in a spectrum such as that of safflower oil (Gunstone, *The Lipid Library*).

From the peak areas of appropriate signals the average number of double bonds per triacylglycerol molecule and the average molecular weight can be calculated and hence the iodine value (excluding unsaponifiable material). These are based on signals at 24.85 (C-3, a measure of total acyl chains), 25.62 (L11, a measure of linoleic acid), 27.15 (O8, O11, L8, L14, monoenes, and dienes), and the multiplet at 29.45 ppm (mid-chain methylenes).

^{31}P NMR spectroscopy is used in the analysis of phospholipids. The phosphorus atom in each phospholipid class (PC, PE, PS, PI, etc.) gives a distinct signal and it is possible to determine the proportion of each phospholipid type using triphenyl phosphate as a standard when quantitative results are needed (Diehl, 2002).

4.11   Mass spectrometry

MS is a procedure used to determine the structure of individual molecules. Originally these had to be isolated by standard methods but it is now more usual to combine the mass spectrometer with GC or HPLC so that individual components of a mixture are separated by chromatography and identified by MS. If the compound is already known then its mass spectrum can be compared with that already reported and contained in a data bank (Christie, *The Lipid Library*). If the compound is novel it should be possible to identify it by application of the basic principles of MS.

When a chromatographic separation precedes MS then it is also desirable to quantify the data so that the proportion of each molecular species is also known. This is usually achieved by measurement of the total ion current but accurate quantification requires calibration with standards or the use of isotopic internal standards. In the combined GC–MS procedures it is also necessary to select derivatives that combine ease and completeness of preparation with good chromatographic properties (satisfactory separation under simple GC or HPLC conditions) and good spectroscopic properties (molecular and/or fragment ions that lead to easy recognition of the molecule). This last may require a selection of the appropriate spectroscopic procedure.

When a molecule is ionised (electronically or chemically) it forms a molecular ion (M^+). This may fragment to give one ionised (A^+) and one unionised (B) particle and a mass spectrometer is a device for producing and examining the charged particles. These are separated according to their mass to charge ratio (m/z, where z is usually one). With high-resolution instruments this value can be measured with such accuracy as to indicate the molecular formula of each ion. The intensity of each peak is related to that of the base peak (largest) which is given a value of 100.

$$M \to M^+ \to A^+ + B$$

Electron ionisation (EI) has been the most widely used ionisation technique in the past. This occurs through an exchange of energy between electrons emitted by a glowing filament (usually at 70 eV) and vaporised sample molecules. Under these conditions the molecular ion usually fragments in one or more ways which can be interpreted in terms of the stability of the various atom-to-atom bonds in the ion.

Chemical ionisation (CI) results from gas phase reactions between a small amount of sample and a large amount of reactant gas (such as methane, ammonia, or isobutene) itself ionised by EI producing reactant gas ions (CH_5^+, NH_4^+, $C_4H_9^+$). CI is usually 'softer' than EI with the consequence that more of the molecular ion is available for detection and fragmentation is less extensive. This generally makes interpretation simpler. The following CI techniques are used by lipid analysts:

- Atmospheric pressure chemical ionisation (APCI)
- Fast atom bombardment (FAB)
- Collision-induced dissociation (tandem mass spectrometry, MS/MS).

For the structural identification of fatty acids, MS procedures linked to GC or HPLC have replaced the older classical methods. MS was first carried out on methyl esters but this is not very satisfactory because under EI the double bonds migrate along the chain and cannot be located unequivocally. Several methods of 'fixing' the double bond were devised but only one of these, applied mainly to monoene esters, remains in use. For both mono- and polyunsaturated acids the methyl esters have been replaced by other acid derivatives that give useful structural information. Appropriate fatty acid derivatives are now generally examined by EI and triacylglycerols by one of the CI methods.

Olefinic esters react with dimethyldisulphide (MeSSMe) and iodine to give a bis(methylthio) derivative the mass spectrum of which shows a molecular ion and two or three large fragment ions that together clearly indicate the position of the SMe groups and hence of the double bond.

$$RCH=CHR' \rightarrow RCH(SMe)CH(SMe)R'$$

For example, methyl oleate gives a molecular ion at 390 ($C_{21}H_{42}O_2S_2$) and three large fragment ions at 173 ($C_9H_{18}SMe$), 217 ($C_{10}H_{18}O_2SMe$), and 185 (217–32 through loss of methanol). These clearly show that methyl oleate is $\Delta9$–18:1 but do not indicate the configuration of the double bond. However, cis and trans monoenes form threo and erythro adducts, respectively and although these have similar mass spectra they are separated by

GC. The procedure is less satisfactory with polyunsaturated acids which are better examined in other ways.

Polyunsaturated acids are now usually examined as picolinyl esters or as 2-alkyl-4,4-dimethyloxazoles (DMOX). These compounds have the structures indicated in Figures 4.4 and 4.5. When these molecules are ionised the charge is carried on the nitrogen atom and double-bond ionisation and isomerisation are minimised. Radical-induced cleavage occurs evenly along the chain and gives a series of relatively abundant ions of high mass resulting from the cleavage of each C−C bond. When a double bond or other functional group is reached then diagnostic ions appear.

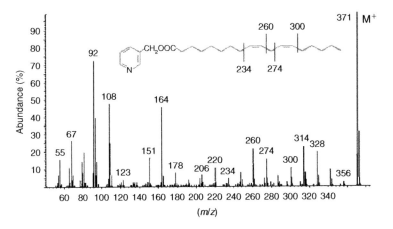

Figure 4.4 Mass spectrum of linoleic acid as the picolinate (*Source*: Downloaded with permission from MS files, www.lipid.co.uk).

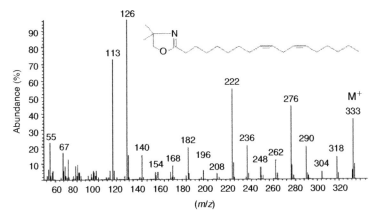

Figure 4.5 Mass spectrum of linoleic acid as the DMOX derivative (*Source*: Downloaded with permission from MS files, www.lipid.co.uk).

The picolinyl esters are made from the free acids and picolinyl alcohol either via the acid chloride (formed by reaction with oxalyl chloride) or through interaction with 1,1'-carbonyldiimidazole in dichloromethane in the presence of 4-pyrrolidinopyridine as catalyst. Another method involves interesterification of triacylglycerols or phospholipids with potassium *tert*-butoxide and 3-hydroxymethylpyridine for 2 min at room temperature.

The spectrum shows some fragments of low mass characteristic of picolinates resulting from $ArCH_2^+$ (93), $ArCH_2O^+$ (108), $ArCH_2OC(OH){=}CH_2^+$ (151), and $ArCH_2OC(O){=}CH_2^+$ (164) where Ar is C_5H_4N or C_5H_5N. In addition there is a molecular ion peak and a series of other high mass fragments which, correctly interpreted, will indicate a structure for the picolinate. For example the ester from linoleic acid (Figure 4.4) has peaks at 371 (M^+, the 18:2 picolinate which is a C_{24} compound), 356 (M^+-15), and a number of fragments with lower m/e values. Most of these ions differ by 14 mass units from their neighbour representing loss of CH_2 but something different happens between 300 and 274 and between 260 and 234 where there is a loss of 26 mass units (C_2H_2 representing a $-CH{=}CH-$ unit). These fragments indicate the presence of double bonds at $\Delta 9$ and $\Delta 12$.

DMOX derivatives are made by heating the lipid with 2-amino-2-methyl-1-propanol in a nitrogen atmosphere at 180°C (2 h for free acids, 18 h for methyl or glycerol esters). Their spectra show peaks at 113 and 126 common to all DMOX derivatives along with a molecular ion and a series of fragments differing by 14 mass units except that some pairs differ by only 12 mass units. The latter are indicative of olefinic centres and are interpreted according to the statement: 'if there is an interval of 12 mass units between the most intense peaks of clusters of ions containing n and $n-1$ carbon atoms then there is a double bond between carbon $n+1$ and n in the acyl chain'. Spectra of DMOX derivatives of many acids are available on Christie's website.

MS procedures, combined with a chromatographic separation system, also give valuable insight into the structure and composition of triacylglycerol mixtures such as milk fats, vegetable oils, and fish oils. In general, identification depends on molecular ions that define the number of both carbon atoms and double bonds in each triacylglycerol molecule. In addition, fragment ions indicate the nature of each acyl group in terms of its number of carbon atoms and unsaturated centres and in some cases will define the distribution of fatty acyl residues between the primary (*sn*-1/3) and secondary (*sn*-2) glycerol positions. Quantitative determination of mixtures is still a problem because the MS responses of triacylglycerols vary with the molecular structure. This topic is intensively reviewed by Laakso and Manninen in Hamilton and Cast (1999) and by Laakso in Dobson (2002).

Reverse phase HPLC followed by APCI MS gives a molecular ion $(M+H)^+$ and fragment ions corresponding to M–RCOOH. For example, the StLO fraction gives peaks at 885.6 $(M+H)^+$, 605.4 $(StO)^+$, 603.5 $(StL)^+$, and 601.4 $(OL)^+$. Among the diacylglycerols, the 1,3 isomer is less intense than the 1,2 and 2,3 isomers and this makes it possible to identify the fatty acid at the 2 position.

Mass spectrometry of lipids has also been reviewed by Christie (1998 and *The Lipid Library*), by Laakso and Manninen and by Roach *et al.* in Hamilton and Cast (1999), and by Dobson and Christie, Laakso, and Korachi *et al.* in Dobson (2002).

Physical Properties

5.1 Polymorphism, crystal structure, and melting point

In the solid state long-chain compounds frequently exist in more than one crystalline form and consequently have more than one melting point. This property (polymorphism) is of both scientific and technical interest. Understanding this phenomenon is essential for the satisfactory blending and tempering of fat-containing materials (such as spreads and confectionery fats) which must attain a certain physical appearance during preparation and maintain it during storage. Problems of graininess in margarine and bloom in chocolate, for example, are both related to polymorphic changes (Sections 8.3 and 8.7). The experimental methods used most extensively to examine melting and crystallisation phenomena involve low-resolution pulsed ^{1}H NMR spectroscopy, differential scanning calorimetry, infrared (IR) spectroscopy, and X-ray diffraction (Larsson *et al.*, 2006).

X-ray investigations indicate that the unit cell for long-chain compounds is a prism with two short spacings and one long spacing as indicated in Figure 5.1. When the long spacing is less than the molecular dimension calculated from known bond lengths and bond angles, it is assumed that the molecule is tilted with respect to its end planes. Sometimes, however, the length is such as to indicate a dimeric or trimeric unit for the most stable form. The molecules assume the angle of tilt at which they are most closely packed. This will give the greatest physical stability and the highest melting point.

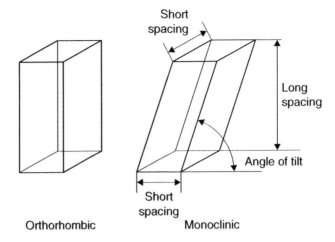

Figure 5.1 The unit cell of long-chain compounds (kindly supplied by my colleague Dr C. M. Scrimgeour).

5.2 Alkanoic and alkenoic acids

The melting points of some long-chain acids and their methyl esters are listed in Table 1.1. These values show alternation with increasing chain length, a phenomenon commonly displayed in the physical properties of long-chain compounds in the solid state and related to the arrangement of molecules in the crystals. The melting points of acids with an even number of carbon atoms in the molecule and their methyl esters plotted against chain length fall on smooth curves lying above similar curves for the odd acids and their methyl esters. Odd acids melt lower than even acids with one less carbon atom. The two curves for saturated acids converge at 120–125°C.

The melting points of unsaturated acids depend on chain length and on the number, position, and configuration of the unsaturated centres. For example stearic (70°C), oleic (Δ9c, 11°C), elaidic (Δ9t, 45°C), and stearolic acids (Δ9a, 46°C) have the melting points shown. Note the considerable difference between *cis* and *trans* isomers. Among polyunsaturated acids those with conjugated unsaturation are higher melting than their methylene-interrupted isomers (Tables 1.1 and 5.1).

Alkanoic acids exist in three polymorphic forms designated A, B, and C for acids with an even number of carbon atoms. Form C has the highest melting point and is the most stable (physically). It is obtained by crystallisation either from the melt or from polar solvents.

Table 5.1 Melting points (°C) of some mono and poly-unsaturated acids

Monoenes	
16:1 (9c)	0.5
18:1 (9c)	16.3
20:1 (9c)	25
22:1 (13c)	33.4
Polyenes with methylene-interrupted unsaturation	
18:2 (9c12c)	−5
18:2 (9c12t)	−3
18:2 (9t12t)	29
18:3 (9c12c15c)	−11
18:3 (9t12t15t)	30
Polyenes with conjugated unsaturation	
18:2 (9c11t)	22
18:2 (9t11t)	54
18:3 (9c11t13c)	44
18:3 (9c11t13t)	49
18:3 (9t11t13c)	32
18:3 (9t11t13t)	71

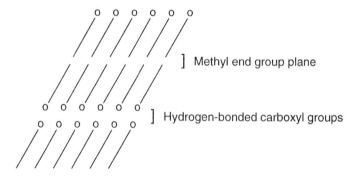

Figure 5.2 Schematic arrangement of alkanoic acid molecules in the crystalline form. The o represents the polar head group (COOH) and the line represents the alkyl chain which will assume a zig-zag arrangement of successive carbon atoms.

Crystallisation from non-polar solvents gives form A or forms B and C. The molecules crystallise in dimeric layers. Alternation of melting point for odd and even chain-length compounds results from the fact that the methyl groups in the end group plane interact differently in the odd and even series (Figure 5.2).

5.3   Glycerol esters

For most technical purposes the melting behaviour of triacylglycerols is more important than that of the fatty acids. It has long been known that fats show multiple melting points and as far back as 1853 glycerol tristearate was reported to have three melting points at 52°C, 64°C, and 70°C. When the melt of a simple triacylglycerol (GA_3) is cooled quickly it solidifies in its lowest melting form (α) with perpendicular alkyl chains in its unit cell (the angle of tilt is 90°). When heated slowly this melts but, held just above this melting point, it will re-solidify in the β' (beta prime) crystalline form. In the same way a still more stable β form can be obtained from the β' form. The β form with the highest melting point is obtained directly by crystallisation from solvent. The β' and β forms have tilted alkyl chains which permit more efficient packing of the triacylglycerols in the crystal lattice. Glycerol esters with only one type of acyl chain are easy to make and have been thoroughly studied. The results have provided useful guidance but such molecules are not generally significant components of natural fats (except after complete hydrogenation). With mixed saturated triacylglycerols such as PStP (P = palmitic, St = stearic) the β form is only obtained with difficulty and such compounds usually exist in their β' form. Among triacylglycerols with saturated and unsaturated acyl chains, symmetrical compounds (SUS and USU) have higher melting (more stable) β forms but the unsymmetrical compounds (USS and UUS) have stable β' forms (S = saturated and U = unsaturated acyl chains) (Table 5.2).

The stable β form generally crystallises in a double chain length arrangement (DCL or β_2) but if one acyl group is very different from the others either in chain length or in degree of unsaturation the

Table 5.2   Characteristics of α, β', and β forms of crystalline triacylglycerols

Form	MP	Short spacings (nm)	IR absorption (cm^{-1})	Hydrocarbon chain	Subcell
α	Lowest	0.4	720	Perpendicular	Orthorhombic
β'	Intermediate	0.42–0.43 and 0.37–0.40	726 and 719	Tilted	Orthorhombic
β	Highest	0.46 and 0.36–0.39	717	Tilted	Triclinic

crystals assume a triple chain length arrangement (TCL or β_3) to allow more efficient packing of alkyl chains and head groups. These crystals have the short spacing expected of a β crystalline form but the long spacing is about 50% longer than usual (Figure 5.3).

In the DCL arrangement the molecules align themselves (like tuning forks or chairs) with two chains in extended line (to give the DCL) and a third parallel to these (Figure 5.3). Some mixed glycerol esters which have a TCL form when crystallised on their own, give high-melting (well-packed) mixed crystals with a second appropriate glycerol ester (e.g. CPC and PCP or OPO and POP where C = capric, P = palmitic, and O = oleic). This has been described as 'compound formation'.

The methyl groups at the top and bottom of each triacylglycerol layer do not usually lie on a straight line, but form a boundary with a structure depending on the lengths of the several acyl groups. This is called the 'methyl terrace'. The molecules tilt with respect to their methyl end planes to give the best fit between the upper methyl terrace of one row of glycerol esters with the lower methyl terrace of the next row of esters. There may be several β_2 modifications differing in the slope of the methyl terrace and in their angle of tilt.

Crystallisation occurs in two stages: nucleation and growth. A crystal nucleus is the smallest crystal that can exist in a solution and is dependent on concentration and temperature. Spontaneous (homogeneous) nucleation rarely occurs in fats. Instead heterogeneous nucleation occurs on solid particles (dust, etc.) or on the walls of the container. Once crystals are formed fragments may drop off

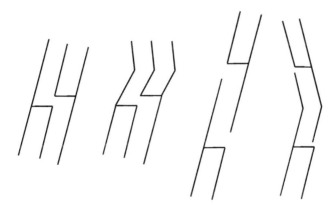

Figure 5.3 DCL and TCL structures (kindly supplied by my colleague Dr C. M. Scrimgeour).

and either re-dissolve or act as nuclei for further crystals. The latter is not desirable in fat crystallisation so agitation during crystallisation should be kept to the minimum required to facilitate heat transfer. Nucleation rates for the different polymorphs are in the order $\alpha > \beta' > \beta$ so that α and β' crystals are more readily formed in the first instance even though the β polymorph is the most stable and is favoured thermodynamically. Crystal nuclei grow by incorporation of other molecules from the adjacent liquid layer at a rate depending on the amount of supercooling and the viscosity of the melt (Timms in Gunstone and Padley, 1997; Sato, 2001; Lawler and Dimick, 2002).

In the production of margarines and shortenings the β' crystalline form is preferred to the β form. β' Crystals are relatively small and can incorporate a large amount of liquid. This gives the product a glossy surface and a smooth texture. β Crystals, on the other hand, though initially small, grow into needle-like agglomerates. These are less able to incorporate liquids and produce a grainy texture. Margarines and shortenings, made from rape/canola, sunflower, or soybean oil after partial hydrogenation, tend to develop β crystals. This can be inhibited or prevented by the incorporation of some hydrogenated palm oil or palm olein which stabilise the crystals in the β' form. These changes in crystallisation pattern are linked with the larger amount of palmitic acid in the palm products. Glycerol esters with C_{16} and C_{18} acyl chains are more likely to be stable in the β' form than glycerol esters with three C_{18} chains.

Because of the importance of its melting behaviour the polymorphism displayed by cocoa butter has been thoroughly investigated. This material is particularly rich in three 2-oleo-1,3-disaturated glycerol esters namely POP, POSt, and StOSt. The solid fat has been identified in six crystalline forms designated I–VI with the melting points and DCL/TCL nature indicated in Table 5.3. Of these, form V (β_2) is the one preferred for chocolate. This crystalline form gives good demoulding characteristics, has a stable gloss, and shows a favourable snap at room temperature. All of these are important properties

Table 5.3 Polymorphism in cocoa butter

	I	II	III	IV	V	VI
MP (°C)	17.3	23.3	25.5	27.3	33.8	36.3
Chain length[*]	D	D	D	D	T	T

[*]D = double chain length, T = triple chain length.

in good chocolate. Two procedures have been employed to promote the formation of this particular crystalline form. The most extensively used is tempering (*i.e.* putting molten chocolate through a series of cooling and heating processes). This optimises the production of the appropriate polymorph. An alternative procedure requires seeding of the molten chocolate with cocoa butter already prepared in form V (β_2) or VI (β_1) but this method is restricted by the difficulty of obtaining adequate supplies of these crystalline forms.

The synthetic glycerol ester 2-oleo-1,3-dibehenin (BOB, O = 18:1, B = 22:0) may be added to cocoa butter to prevent bloom formation by keeping it in its form V at temperatures above 30°C (Section 8.7).

Oils rich in saturated acids contain high-melting triacylglycerols that may crystallise from the oil when stored. When this is considered to be undesirable the oil is subjected to winterisation. The oil is chilled gradually and kept at around 5° for several hours before being filtered. The liquid fraction should then remain clear at ambient temperature. This process is applied to cottonseed oil and to partially hydrogenated soybean oil.

Timms (1978) reviewed and significantly extended information on the heats of fusion of glycerides. He derived an equation for the heat of fusion of mono acid glycerides in the β polymorph form and showed how this could be adapted to calculate the heat of fusion of most glycerides of commercial interest.

5.4   Ultraviolet spectroscopy

The use of ultraviolet (UV) spectroscopy in the study of lipids is confined to systems containing or generating conjugated unsaturation. It is therefore of value in the study of the rare natural acids with conjugated unsaturation. Conjugated dienes such as CLA have a UV maximum around 230–240 nm and trienes show triple peaks around 261, 271, and 281 nm. The more common methylene-interrupted polyenes do not show any interesting UV absorption until double bonds migrate to form conjugated systems. This happens during autoxidation (Section 6.2), alkali isomerisation, and other reactions involving doubly allylic methylene groups. UV spectroscopy is also used in the study of carotenoids with extended conjugated systems (Young and Hamilton in Hamilton and Cast, 1999; Angioni *et al.*, in Dobson, 2002).

5.5   IR and Raman spectroscopy

IR spectroscopy has been applied to solid lipids to provide information about polymorphism, crystal structure, conformation, and chain length but the commonest use of traditional IR spectroscopy has been the recognition and measurement of *trans* unsaturation in acids and esters where unsaturation is normally predominantly *cis*, using neat liquids or solutions. One *trans* double bond absorbs at $968\,cm^{-1}$. Additional *trans* centres increase the intensity but do not change the frequency unless they are conjugated when small changes are reported.

There is no similar diagnostic absorption for a *cis* olefin but Raman spectra show strong absorption bands at $1665 \pm 1\,cm^{-1}$ (*cis* olefin), $1670 \pm 1\,cm^{-1}$ (*trans* olefin), and $2230 \pm 1\,cm^{-1}$ and $2291 \pm 2\,cm^{-1}$ (acetylenes). Carbonyl compounds have a strong absorption band in the region $1650-1750\,cm^{-1}$. This wavelength varies slightly with the nature of the carbonyl compound as in the following saturated and $\alpha\beta$-unsaturated compounds, respectively: aldehydes ($1740-1720$ and $1705-1680\,cm^{-1}$), ketones ($1725-1705$ and $1685-1665\,cm^{-1}$), acids ($1725-1700$ and $1715-1690\,cm^{-1}$), and esters ($1750-1730$ and $1730-1715\,cm^{-1}$).

The analytical uses of FTIR and of NIR have been discussed in Section 4.8.

5.6   Nuclear magnetic resonance spectroscopy

See Sections 4.9 and 4.10.

5.7   Mass spectrometry

See Sections 4.11.

5.8   Density

Density may not seem an exciting physical property to many technologists but it is very important in the trading of oils since shipments are sold on a weight basis but measured on a volume basis. These two values are related by density so it is important to have correct and agreed values for this unit. This is not the same for all oils. It depends on fatty acid composition and minor components as well as on the temperature. Pantzaris has derived an equation taking these variables into account. It is based on iodine value, saponification value, and temperature.

$$d = 0.8543 + 0.000308(SV) + 0.000157(IV) - 0.00068t$$

Where d = apparent density (g/ml or kg/l), SV = saponification value, IV = iodine value, and t = temperature (°C).

Density can be defined in various ways and the correct form must be used when relating volume to weight.

- Density (absolute density or density in vacuum) is: *Mass in vacuum of a volume of oil at t°C ÷ volume of the oil at the same temperature expressed in g/ml or kg/l.*
- Apparent density (density in air, weight-by-volume, or litre-mass) is: *Mass in air of a volume of oil at t°C ÷ volume of the oil at the same temperature expressed in g/ml or kg/l.*
- Relative density (specific gravity, density in relation to water) is: *Mass in air of a given volume of oil at t_1°C ÷ mass in air of same volume of water at t_2°C.* This is a ratio without units. It is important to note that two temperatures are involved and the value is meaningless unless both figures are cited. This is the value most usually employed and equations exist to connect these three expressions.

Further information is given by Gunstone in Hamm and Hamilton (2000).

5.9 Viscosity

Viscosity can be reported as kinematic viscosity or dynamic viscosity with the two values being related through density. The viscosity of a vegetable oil depends on its chemical composition (summarised in the IV and SV) and the temperature of measurement. Equations have been derived which permit calculation of viscosity from knowledge of the other parameters. These have been developed empirically from observations with a range of oils at different temperatures. Viscosity has been correlated with density, refraction, surface tension, and other physical properties and the relation between temperature and viscosity for selected oils has been described. Detailed references for these matters can be found in Gunstone (2004).

5.10 Refractive index

The refractive index is easily measured on small amounts of material. Refractive index increases with chain length (though not in a linear fashion) and with increasing unsaturation. Geometric isomers differ from one another and methylene-interrupted polyenes differ from those with conjugated unsaturation. Triacylglycerols have higher values than free acids. Values for commercial oils are cited in Table 5.4.

5.11 Solubility of gases in oils

A discussion (Hilder in Gunstone and Padley, 1997) on the solubility of gases in oils includes the data presented in Tables 5.5 and 5.6 for oxygen, nitrogen, and air. When an oil is in contact with air the dissolved gases will depend on their individual solubility as well as their concentration in air. The high solubility of the monatomic argon enhances its concentration so that 1% in air becomes 3% of the gases in the oil. The solubility of hydrogen in oils and fats is discussed in Section 6.2.

5.12 Other physical properties

Gross heats of combustion (HG) for saturated and unsaturated triacylglycerols can be related to the number of valence electrons (EN). The following equations have been derived:

$$HG = -109.20 + 26.39 \text{ EN} \quad \text{saturated triacylglycerols}$$
$$HG = 115.87 + 25.88 \text{ EN} \quad \text{unsaturated triacylglycerols}$$
$$HG = 1{,}896{,}000/SN - 0.6 \text{ IV} - 1600$$

Table 5.4 Physical and chemical properties of selected commodity oils and fats

	Specific gravity (temperature °C)	Refractive index (40°C)	Refractive index (25°C)	IV	SV	Titre (°C)	Unsaponifiable (%)	MP (°C)
Cocoa butter	0.973–0.980 (25/25)	1.456–1.458	–	32–40	192–200	45–50	0.2–1.0	31–35
Coconut	0.908–0.921 (40/20)	1.448–1.450	–	6–11	248–265	–	<1.5	23–26
Corn	0.917–0.925 (20/20)	1.465–1.468	1.470–1.473	107–128	187–195	–	1–3	–
Cottonseed	0.918–0.926 (20/20)	1.458–1.466	–	100–115	189–198	–	<2	–
Linseed	0.930–0.936 (15.5/15.5)^d	1.472–1.475	1.477–1.482	170–203	188–196	19–21	0.1–2.0	–
Olive	0.910–0.916 (20/20)	–	1.468–1.471	75–94	184–196	–	1.5	–3–0
Palm kernel	0.899–0.914 (40/20)	1.452–1.488	–	14–21	230–254	–	<1.1	24–26
Palm	0.891–0.899 (50/20)	1.449–1.455^e	–	50–55	190–209	–	<1.4	33–40
Palm olein	0.899–0.920 (40/20)	1.459–1.459	–	>55	194–202	–	<1.4	–
Palm stearin	0.881–0.891 (60/20)	1.447–1.451	–	<49	193–205	–	<1.0	–
Peanut	0.914–0.917 (20/20)	1.460–1.465	–	86–107	187–196	–	<1.1	–
Rape^a	0.910–0.920 (20/20)	1.465–1.469	–	94–120	168–181	–	<0.21^f	–
Rape^b	0.914–0.920 (20/20)	1.465–1.467	–	110–126	182–193	–	<0.21^f	–
Sesame	0.915–0.923 (20/20)	1.465–1.469	–	104–120	187–195	–	<2.1	–
Soybean	0.919–0.925 (20/20)	1.466–1.470	–	124–139	189–195	–	1.6	–
Sunflower	0.918–0.923 (20/20)	1.467–1.469	1.472–1.476	118–145	188–194	–	<1.6 (max 2.0)	–
Sunflower^c	0.915–0.920 (20/20)	–	1.467–1.469	75–90			0.8–1.0 (max 2.0)	–

Notes
^{a}High-erucic rape seed oil.
^{b}Low-erucic rape seed oil.
^{c}High oleic sunflower seed oil.
^{d}Also 0.924–0.930 (25/25).
^{e}50°C.
^{f}These values are correctly copied from the source but they are in error. Better values are 0.5–1.2%.
Source: Physical and Chemical Characteristics of Oils, Fats, and Waxes (Firestone, 1999).

CHAPTER 5

Table 5.5 Solubility of oxygen and nitrogen (ppm, 1 bar) in oils

Temperature (°C)	Oxygen	Nitrogen
0	170	80
25	180	85
50	185	90
75	190	95
100	200	105
125	*	110
150	*	115

Source: Hilder in Gunstone and Padley (1997).
*Oxygen solubilities at higher temperatures are not reliable because oxidation occurs.

Table 5.6 Gas content of oil saturated with air

	Solubility (ppm)	Air dissolved in oil (ppm)
Oxygen	180	38
Nitrogen	85	66
Argon	270	3

Source: Hilder in Gunstone and Padley (1997).

In a useful paper Coupland and McClements (1997) reported several physical properties (density, viscosity, adiabatic expansion coefficient, thermal conductivity, specific heat, ultrasonic velocity, and ultrasonic attenuation coefficient) for a number of liquid oils. Chumpitaz *et al.* (1999) recently reported the surface tension of several fatty acids and triacylglycerols.

These data are important for processes involving gas–liquid contact such as distillation and stripping columns, deodorisers, reactors, and equipment for physical refining.

Useful data taken from the AOCS publication 'Physical and Chemical Characteristics of Oils, Fats, and Waxes' is given in Table 5.5.

Chemical Properties

This chapter covers the chemical reactions that are significant in the food industry and include hydrogenation, oxidation, thermal changes in double bond systems, and some reactions of the acid/ester function.

6.1 Hydrogenation

As generally practised hydrogenation involves reaction between the unsaturated centres in an oil or fat in the presence of a metallic catalyst. This is a heterogeneous reaction, involving solid, liquid, and gas, taking place on the solid catalyst surface at an appropriate temperature and pressure. After hydrogenation, the product has changed physical, chemical, and nutritional properties. Sometimes compromises have to be made between these.

Each year millions of tonnes of soybean and other unsaturated vegetable oils containing oleic, linoleic, and linolenic acids as well as diminishing levels of fish oils with more complex patterns of unsaturation are subject to hydrogenation. This reaction is practised in three major ways:

(1) Brush hydrogenation is a short reaction, applied particularly to soybean and rapeseed/canola oils, designed to reduce the level of linolenic esters (18:3) to around 4%, thereby increasing shelf life. This is not a very extensive hydrogenation process. Apart from the lowering of the level of linolenate the fatty acids are little changed and production of *trans* isomers will be minimal. The partially reduced linolenate will be converted not to linoleate but to a mixture of 18:2 isomers and

perhaps some 18:1 (not oleate). Table 2.5 contains data on the composition of various products when soybean oil of iodine value (IV) 132 is hydrogenated progressively to oil of IV 110 (brush hydrogenation), 97, 81, and 65.

(2) Partial hydrogenation is an important way of processing oils and fats to extend their range of use. The technique, first applied to such materials by the German chemist Normann, has been in operation for over 100 years and has been subject to continuous improvement during that time. The main objective is to convert a liquid oil (vegetable or fish) into a semi-solid fat that can be used as a component of a spread. Compared to brush hydrogenation this is a more extensive reaction whereby the content of polyunsaturated fatty acid is much reduced and a considerable proportion of *trans* 18:1 is formed. This results in a rise in melting point (because *trans* esters are higher melting than their *cis* isomers) that affects spreadability, oral response, and baking performance. Two other changes have also to be considered. There is an increase in oxidative stability through the complete or partial removal of the polyunsaturated fatty acids that are so easily oxidised and there is a decrease in nutritional value through the destruction of essential fatty acids and the formation of saturated acids and of unsaturated acids with *trans* configuration. During hydrogenation linoleate is reduced first to a mixture of *cis* and *trans* 18:1 isomers referred to as 'oleate' and then to stearate. The level of stearic acid may also increase (Table 2.5) but it has been argued that this is not a serious cholesterol-raising fatty acid. Along with ruminant fats, partially hydrogenated vegetable oils are the most important source of *trans* acids and, following the concern over these acids, there is now a requirement in some countries to indicate their level on the product label. As a consequence other ways to achieve the desired physical properties have been explored. These include modifications of the hydrogenation process to give less *trans* compounds and using interesterification of suitable blends as an alternative approach. Concern over *trans* acids is based on the fact that they raise LDL (low-density lipoproteins) levels and lower HDL (high-density lipoproteins) levels. This has been reported to be mainly a US problem because 75% of the fats consumed in that country are derived from soybean oil with its high

levels of polyunsaturated fatty acid. According to US legislation samples containing less than 0.5 g of non-conjugated
trans acids per 14 g serving may be claimed as zero *trans*.

(3)   Complete hydrogenation is a still more extensive process in
which virtually all unsaturated acids are converted to their
saturated analogues. This produces 'hardstock' which can
be blended with a liquid (unsaturated) oil and subsequently
interesterified. Hardstock is rich in stearic acid formed from
the unsaturated C_{18} acids originally present. It usually has an
IV of around 2. It contains only low levels of unsaturated acid
and therefore very little *trans* acids. However, it may be necessary for the word 'hydrogenated' to appear on the label.

At the molecular level one or more of the following changes may
occur during this reaction: hydrogenation (saturation) of unsaturated centres, stereomutation of natural *cis* olefins to their higher-
melting *trans* isomers, double bond migration, and conversion of
polyunsaturated fatty acids to monounsaturated and saturated
acids. These are the consequences of reaction between a liquid
(fatty oil) and a gas (hydrogen) occurring at a solid surface (the
catalyst). In the sequence below the horizontal line shows the conversion of diene to monoene and of monoene to saturated acid/
ester via the half hydrogenated states represented as DH and MH.
The steps shown vertically are the reverse processes whereby DH
reverts to D or a diene isomer and MH reverts to M or a monoene
isomer. It is during these reverse stages that *trans* and positional
isomers are formed. There are six stages altogether and it is important to understand the relative rates of these. In the conversion of
D to M the first step is rate-determining and the second step is
fast. Levels of DH will therefore be low and the conversion of DH
back to D is slow and only important in the unusual situation that
hydrogen is present in very low concentration. In the conversion of
M to S the final stage is slow and rate-determining. This makes it
more likely that there will be considerable recycling of M and MH
leading to formation of stereochemical and positional isomers.

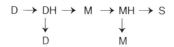

The catalyst used on a commercial scale is nickel on an inert sup-
port at a 17–25% level, encased in hardened fat. This preserves

the activity of the nickel in a form that is easily and safely handled. Hydrogenation is generally conducted at 180–200°C and 3 bar pressure in vessels containing up to 30 tonnes of oil. To minimise the use of catalyst it is desirable to use refined oil and the highest quality of hydrogen. Through improvements in the quality of catalyst and in equipment the requirement for catalyst has been gradually reduced. In 1960, 0.2% nickel was required but by the end of the century this was reduced 4- to 8-fold to between 0.025% and 0.05%. Reaction may proceed in a batch-wise manner with up to 8–10 batches in a 24-hour day or in a semi-continuous fashion at rate of 25–100 tonnes per day. The reaction is exothermic and appropriate cooling is required as well as stirring to distribute the heat.

Several significant variables have to be considered:

- The nature of the oil being treated.
- The extent of hydrogenation which is desired.
- The selectivity to be achieved in terms of PUFA–MUFA-saturated ratios and the ratio of *cis* to *trans* isomers.
- The quality and quantity of catalyst in terms of pore length, pore diameter, activity, and amount used.
- The reaction conditions of temperature, pressure, and degree of agitation.

The competition between hydrogenation (a change incorporating hydrogen) and isomerisation (a change not involving additional hydrogen in the product) depends on the availability of hydrogen at the catalyst surface in relation to the demand. A plentiful supply of hydrogen will promote hydrogenation, and an inadequate supply of hydrogen will allow isomerisation to become more significant. The availability of hydrogen at the catalyst surface is enhanced by increased pressure and increased agitation. The demand for hydrogen is increased with higher temperatures, higher catalyst quantity, higher catalyst activity, and more highly unsaturated oils.

The progress of the reaction can be followed in a number of ways that vary in simplicity, in speed of completion, and in the information they provide. They include: the volume of hydrogen used which will measure saturation but not isomerisation, iodine value measured by an accelerated technique that will provide similar information, refractive index, solid fat content measured by low-resolution ^{1}H-NMR, solid fat index measured by dilatometry, slip melting point, or gas chromatography of methyl esters.

Koetsier in Gunstone and Padley (1997) has summarised data on the solubility of hydrogen in vegetable oil. This information is obviously important for hydrogenation. He cites solubility values (maximum concentration in oil at a given temperature and pressure) from two sources at 1 bar and 100–200°C of 2.60–3.36 and 2.76–3.40 mol/m^3. The concentration of hydrogen is thus much lower than the concentration of unsaturated centres and for a fish oil of iodine value hydrogenated at 5 bar and 180°C Koetsier gives concentrations of ~7000 and 16 mol/m^3, respectively, for the olefinic groups and the hydrogen.

6.2   Atmospheric oxidation

Unsaturated fats like other unsaturated products such as rubber and paints deteriorate as a consequence of reaction with oxygen (air) that leads, in fat-containing foods, to oxidative rancidity (another form of rancidity results from hydrolysis). Oxidative deterioration is of two kinds (autoxidation and photo-oxidation). These processes lead first to similar, but not identical, unstable allylic hydroperoxides which decompose to volatile short-chain molecules (mainly aldehydes) responsible for the undesirable odours and flavours associated with oxidative rancidity (Figure 6.1). These compounds have low but differing threshold levels so that only small quantities are necessary to produce their undesirable effects. Since the oxidation processes are influenced by heat, light, the presence of pro-oxidants (copper and iron) and antioxidants (natural or synthetic), the presence of already-oxidised material, and of air, attention must be given

Olefinic acids/esters   (methyl oleate, linoleate, etc., glycerol esters, oils and fats)
↓
Allylic hydroperoxides   (highly reactive species)
↓
- Volatile compounds of lower molecular weight (aldehydes, etc.) which provide odour and flavour, often at low concentration.
- Compounds with the same chain length: rearrangement products, products of further oxidation, and products of reaction with other components in the reaction system.
- Compounds of higher molecular weight such as dimers and polymers.

Figure 6.1   Formation and further reactions of allylic hydroperoxides.

to all these factors in the handling, transport, and storage of oils and fats and of foods containing these.

The level and nature of unsaturation is an important factor in the rate of oxidation (Table 6.1). This is the reason for the instability of (highly unsaturated) fish oils and accounts for the short shelf life of oils containing linolenic acid (soybean and rapeseed/canola oils) and the preference for oils with reduced levels of this acid available through breeding or through brush hydrogenation. The figures in Table 6.1 indicate that the polyunsaturated acids (linoleic, linolenic, etc.) oxidise much quicker than the monounsaturated oleic acid. Oxidative deterioration is strongly linked with the number of doubly activated allylic groups present in PUFA. An index of the oxidisability of vegetable and fish oils has been expressed as:

$$\text{Oxidisability} = (\text{diene}) + 2(\text{triene}) + 3(\text{tetraene}) + 4(\text{pentaene}) + 5(\text{hexaene})$$

It is not possible to prevent these oxidation reactions but they can be inhibited and the induction period (Section 4.6) should be extended as much as possible. The difficulties of inhibiting oxidation are further complicated by the fact that many foods (and biological systems) are emulsions of lipids and aqueous systems and attention has to be given to the distribution of antioxidants and pro-oxidants between these two phases and at the interface between them.

The detailed structure of the hydroperoxides that can result from oleate, linoleate, or other polyunsaturated acid is important as this controls the chemical structure of the volatile short-chain compounds, each of which has its own flavour/odour and its own threshold level. One of the problems with linolenic esters is that trienes oxidise quicker than dienes and the resulting volatile aldehydes have lower threshold values. When an oxidised fatty molecule cleaves one fragment will remain attached to glycerol (the core aldehyde) while the methyl end of the molecule will provide

Table 6.1 Relative rates of autoxidation and photo-oxidation of oleate, linoleate, and linolenate (autoxidation of methyl oleate = 1)

Reaction	Oxygen	18:1	18:2	18:3
Autoxidation	Triplet	1	27	77
Photo-oxidation	Singlet	3×10^4	4×10^4	7×10^4
Ratio of reaction rates		30,000	1500	900

the volatile component. The latter can be removed through refining but the short-chain fragment still attached to glycerol will probably remain in the oil. The sequence in Figure 6.2 shows the formation of octanal from a glycerol ester containing oleic acid via the 11-hydroperoxide. Other aldehydes from oxidised oleate, linoleate, and linolenate are listed in Table 6.2.

Autoxidation is a radical chain process. That means that the intermediates are radicals (odd electron species) and that like other chain processes there are three stages: initiation, propagation, and termination (Figure 6.2). The initiation step (not fully understood) is followed by a propagation sequence that continues, perhaps for many cycles, until stopped by one of the termination processes. The process will be accelerated by more initiation steps (involving metal ions, higher temperatures, or a poor sample containing already oxidised oil) and by less termination steps resulting in more of the propagation cycle. More important, the process will be inhibited by having fewer initiation steps or more termination steps thereby having fewer and shorter propagation cycles. This can be achieved by starting with good quality oil and by the

Table 6.2 The major hydroperoxides produced from oleate, linoleate, and linolenate during autoxidation and photo-oxidation and the volatile aldehydes resulting from these

Ester	Hydroperoxide	Double bond	Volatile aldehyde*
Oleate	8	9	11:1 (2)
	9	10	10:1 (2)
	10	8	9:0
	11	9	8:0
Linoleate	9	10,12	10:2 (2,4)
	10	8,12	9:1 (3)
	12	9,13	7:1 (2)
	13	9,11	6:0
Linolenate	9	10,12,15	10:3 (2,4,7)
	10	8,12,15	9:2 (3,6)
	12	9,13,15	7:2 (2,4)
	13	9,11,15	6:1 (3)
	15	9,12,16	4:1 (3)
	16	9,12,14	3:0

* Numbers before the colon indicate the number of carbon atoms in each aldehyde molecule and numbers after the colon indicate the number of unsaturated centres. Numbers in brackets indicate double bond position with respect to the aldehyde function.

Glyc-OCO(CH$_2$)$_7$CH=CH(CH$_2$)$_7$CH$_3$ Glycerol oleate
↓
Glyc-OCO(CH$_2$)$_7$CH=CHCH(OOH)(CH$_2$)$_6$CH$_3$ Allylic hydroperoxide
↓
Glyc-OCO(CH$_2$)$_8$CHO + OHC(CH$_2$)$_6$CH$_3$ Core aldehyde and volatile aldehyde

Figure 6.2 Formation of a typical core aldehyde and a short-chain volatile aldehyde (octanal) from a glycerol ester through the appropriate allylic hydroperoxide. Other aldehydes result from other hydroperoxides (Table 6.2).

Initiation	RH → R•	Resonance-stabilised alkyl radical
Propagation	R• + O$_2$ → RO$_2$•	Fast reaction to give a peroxy radical
	RO$_2$• + RH → RO$_2$H + R•	Rate-determining step
Termination	RO$_2$• + RO$_2$• → stable products	
	RO$_2$• + R• → stable products	
	R• + R• → stable products	

Figure 6.3 Olefin autoxidation. RH represents an olefinic compound in which H is attached to an allylic carbon atom. RO$_2$H is a hydroperoxide.

presence of appropriate antioxidants (see below). The initiation and propagation steps involve breaking a C−H in the glycerol ester. The energy required to remove hydrogen from a saturated methylene group, an allylic methylene group, and a doubly allylic methylene group (as at C11 in linoleate) is 100, 75, and 50 kcal, respectively. These values relate to the relative ease of oxidation of saturated, oleate, and linoleate (Table 6.1). The allyl radical first produced is resonance stabilised so that the radical centre is spread over different carbon atoms giving rise to the various hydroperoxides listed in Table 6.2. In the oxidation of olefinic lipids there is normally an induction period during which reaction is very slow and deterioration is not significant, followed by a quicker stage of undesirable oxidation. One purpose of antioxidants is to extend this induction period.

Sterols are also subject to oxidation. Cholesterol (Section 1.5) contains a cyclic double bond (Δ5) and two tertiary carbon atoms in its side chain (C-20 and C-25), all sites where oxidation may occur. Some cholesterol oxides are produced as part of the normal metabolism of cholesterol to bile acids. At higher levels these affect human health by contributing to the development of atherosclerosis. Oxidised animal-based foods represent a primary source

$$cis\text{-}RCH=CHCH_2R' + {}^1O_2 \rightarrow trans\text{-}RCH(OOH)CH=CHR$$

Figure 6.4 Reaction of olefin with singlet oxygen to give allylic hydroperoxides with double bonds in a different position and of changed configuration.

of oxidised cholesterol. Such products are not present in fresh foods but are formed during handling prior to consumption, mainly through autoxidation. Cholesterol esters are predominantly of linoleic acid while free cholesterol is associated with polyunsaturated fatty acids in phospholipids in cell membranes. In both cases oxidation can be initiated in the polyunsaturated fatty acids and then involve the cholesterol molecule. This holds in the animal (before being prepared as food) and in the human, and in both oxidation can be retarded by appropriate dietary antioxidants. Processing conditions should also be adapted to minimise oxidation as, for example, in the preparation of spray-dried eggs. Between 0.5% and 1.0% of dietary cholesterol may be oxidised and the primary oxidation products include 7-α-hydroxy-, 7-β-hydroxy, and 7-keto-cholesterol, cholesterol α- and β-epoxides, 3,5,6-trihydroxycholesterol, and 20- and 25-hydroxycholesterol (Cuppett, 2003).

Photo-oxidation is a quicker reaction between olefin and a light-activated form of oxygen. The activation process requires a sensitiser such as chlorophyll, riboflavin, myoglobin, erythrosine, rose bengal, or methylene blue. The sensitiser absorbs energy from a photon and this energy is eventually passed to oxygen, raising it from the triplet to the more reactive singlet state. Singlet oxygen reacts rapidly with double bonds by an ene reaction to give an allylic hydroperoxide. Photo-oxidation differs from autoxidation in that it is faster (Table 6.1) and its rate is related to the number of double bonds rather than to the number of doubly allylic functions. It is inhibited by appropriate quencher molecules like carotene rather than by the range of compounds that inhibit autoxidation.

Antioxidants are materials present in oils and fats to protect them against autoxidation. They may be natural compounds already present in the oils such as tocols (Section 1.6) and ferulic acid esters (structure below) or they may be natural or synthetic compounds added by the technologist. There is not enough natural antioxidant to meet demand so synthetic compounds must be used in some cases. Some of the natural antioxidants may be lost during refining so that refined oils are generally less stable than crude oils. The eight different tocols have differing antioxidant activity so that total tocol content is not an adequate measure of antioxidant activity. (Table 6.3 and Section 1.5) Antioxidants may also be lost during food preparation so prepared foods may be less stable than the ingredients from which they are made. Autoxidation can be inhibited (induction period extended and shelf life lengthened) but it cannot be reversed so antioxidants should be added as early as possible after the refining process. The amount of antioxidant employed must be the optimum. There are legal limits for the synthetic compounds and tocopherols may act as pro-oxidants at higher concentrations. In the case of emulsions, as opposed to bulk oils, it is important to consider the distribution of antioxidants and pro-oxidants between the oil and water phases. To avoid the reaction of existing hydroperoxides with metal ions which generates more radical species, it is important to keep hydroperoxides and metal ions apart and this may be assisted by controlling the pH of the emulsion and by selection of appropriate emulsifiers.

Some plants have other natural antioxidants in their leaves or seeds. Familiar examples include oat oil (with α-tocopherol,

Table 6.3 Vitamin E content (mg/100 g) of some vegetable oils and of butter and lard

Oil	Tocopherols					Tocotrienols					Grand total	IU
	α	β	γ	δ	Total	α	β	γ	δ	Total		
Soybean	10	–	59	26	96					0	96	24
Corn	11	5	60	2	78					0	78	20
Rapeseed	17		35	1	53					0	53	30
Sunflower	49		5	1	55					0	55	73
Groundnut	13		22	2	37					0	37	23
Cottonseed	39		39		78					0	78	64
Safflower	37		17	24	80					0	80	61
Palm	26		32	7	65	14	3	29	7	53	118	49
Coconut	Trace		Trace		1	Trace		2	Trace	3	4	1
Olive	20	1	1		22					0	22	30
Wheat germ	121	65	24	25	235	2	17			19	254	233
Rice	12	4	5		21	18	2	57		77	98	30
Butter	2				2						2	3
Lard	1				1	1				1	2	2

IU represents total vitamin E content calculated on a weighted basis for the effect of each tocol.
Source: Adapted from Stone and Papas in Gunstone (2003). See also Table 1.3.

α-tocotrienol, and avenathramides), sesame oil (with sesamin, sesamolin, and sesaminol, all of which are derivatives of sesamol – 3,4-methylenedioxyphenol), and ricebran oil (with tocotrienols, avenasterols, and oryzanols which are sterol esters of ferulic acid).

$$4\text{-OH-3-OMeC}_6\text{H}_3\text{CH}=\text{CHCOOH ferulic acid}$$

Antioxidants are also present in herbs and spices and while these can sometimes be used as extracts their food use is limited by strong flavours that may or may not be acceptable in other foods. Tea leaves are a rich source of antioxidants (catechins) as are many fruits and vegetables containing flavonoids. Dietary consumption of these as whole foods provides a good source of the antioxidants required by the body to counter oxidative damage to protein and to DNA caused by radicals produced through lipid oxidation. This applies also to vegetables containing carotenes. Rosemary leaves contain powerful antioxidants such as carnosic acid, carnosol, and rosmarinic acid and rosemary extracts are available for use as antioxidants. Vitamin C (ascorbic acid) acts as an oxygen scavenger, removing traces of residual oxygen in a packed and sealed product. It is water-soluble but can be used in a lipid-soluble form as ascorbyl palmitate. Phospholipids show ill-defined antioxidant activity possibly through activity as a chelating agent and/or emulsifier.

The supply of natural antioxidants is insufficient to meet demand so some use of synthetic antioxidants is obligatory. Even so-called natural vitamin E may have been submitted to a chemical reaction (permethylation) in which tocols with only one or two methyl groups have been converted to their trimethyl derivative (α-tocopherol, Figure 1.5).

The synthetic compounds that can be used as antioxidants in food are strictly controlled, as is the level at which they may be used. The matter is complicated in that not all countries have agreed to the same list of acceptable compounds. This becomes important for materials that are traded between countries having different permitted antioxidants. Obviously the antioxidants must be non-toxic and that must apply also to the products produced from them as a result of their antioxidant activity.

The four important synthetic antioxidants discussed here are solid compounds and may be conveniently used as solutions in propylene glycol, monoacylglycerols, or vegetable oils. They are mono

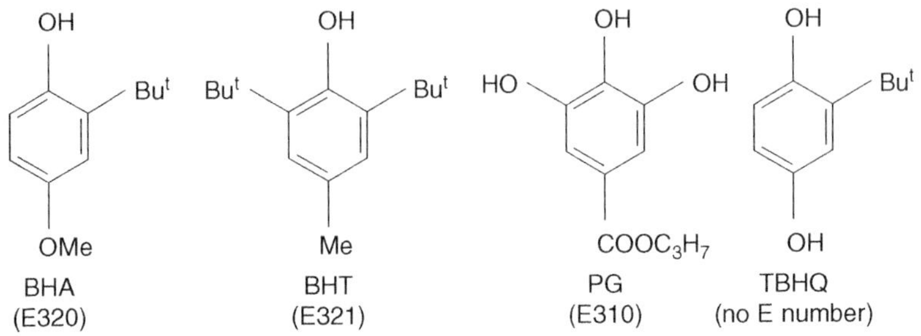

Figure 6.5 The structures and E numbers of synthetic antioxidants. TBHQ has no E number because it is not a permitted antioxidant in the EU.

or dihydric phenols (represented as ArOH) and react readily with a peroxy radical to give a phenoxy radical (ArO·) stabilised by extensive delocalisation of the odd electron over the aromatic system. The E numbers indicate that they may be used in Europe within prescribed limits (Figure 6.5).

Butylated hydroxyanisole (BHA, E320) shows good solubility in fat and reasonable stability in fried and baked products. It is very effective with animal fats and less so with vegetable oils. It shows marked synergism with BHT (butylated hydroxytoluene) and PG (propyl gallate) and can be used at a maximum level of 200 ppm.

BHT, E321 is less soluble than BHA and is not soluble in the propylene glycol frequently used as a solvent for antioxidants. It is synergistic with BHA but not with PG and can be used to a maximum level of 200 ppm.

Synergism is the term used to describe the observation where the efficacy of two or more components is greater than the sum of the effects for individual components and indicates some co-operative activity.

PG, E310 is less soluble than BHA or BHT. It does not generally survive cooking as it decomposes at 148°C. Nevertheless it is effective when used with BHA and may be used up to 100 ppm.

Tertbutyl hydroquinone (TBHQ) is acceptable in USA and many other countries but not in EU-27 and hence does not have an E number. It is very effective with vegetable oils, has good solubility, and is stable at high temperatures. It is frequently used during oil transport and storage and can be subsequently removed during deodorisation.

6.3   Thermal changes

Unsaturated centres may undergo undesirable changes when heated. This is particularly important when the fatty acyl chains contain three or more methylene interrupted olefinic centres and when the temperature exceeds 180°C. This problem may arise during deodorisation of soybean and rapeseed oils containing up to 10% of linolenic acid (18:3) or of fish oils with eicosapentaenoic acid (EPA) (20:5) and docosahexaenoic acid (DHA) (22:6). At elevated temperatures the risk of oxidative change is enhanced but even in the absence of air undesirable changes occur at elevated temperatures. Studies on the refining of vegetable oils have shown that stereoisomerism of linolenic acid with three double bonds occurs from 220°C upwards and is quicker than for linoleate with only two double bonds. The changes that may occur during exposure of polyunsaturated fatty acids to higher temperatures include:

- cyclisation (formation of five- and six-membered carbocyclic ring systems);
- geometrical isomerism (conversion of the natural all-*cis* polyunsaturated fatty acid to isomers with both *cis* and *trans* double bonds);
- polymerisation (producing dimers, trimers, and oligomers of enhanced molecular weight).

All these changes are undesirable on nutritional grounds since the nutritional value of the polyunsaturated fatty acid depends on them retaining their all-*cis* pattern of unsaturation. Special analytical methods may be required to detect these changes.

6.4   Reactions of the carboxyl/ester function

Oils and fats are glycerol esters and the ester functions are reactive centres which can be modified in various ways. For example, they can be hydrolysed. The suffix -lysis means splitting so that hydrolysis implies splitting with water and the final products are fatty acids and glycerol. This reaction generally requires a catalyst which may be acidic, basic, or enzymatic. When catalysed by a lipase the reaction may be described as lipolysis. Hydrolysis is an important reaction in the oleochemical industry but in the food industry the conversion of glycerol esters to other esters is more important.

Esters can be made from fatty acids and alcohols but it is often more convenient to transform existing esters (such as triacylglycerols) to other esters by reaction with an alcohol (alcoholysis), an acid (acidolysis), or another ester (interesterification) using a catalyst that may be chemical or a lipase.

Two important alcoholysis procedures are methanolysis and glycerolysis. The former is used in the conversion of glycerol esters to methyl esters and is used on a mg scale for analytical purposes (gas chromatography) and on a multi-tonne scale to produce esters for biofuel or as an intermediate in the production of fatty alcohols. Glycerolysis involves reaction of triacylglycerols with glycerol to produce monoacylglycerols and diacylglycerols. The former, as such or after further modification, are much used in the food industry as emulsifying agents (Section 8.11).

Acidolysis is less commonly employed but it may be used to incorporate (say) lauric acid (12:0) into a $C_{16/18}$ oil.

Interesterification can be carried out on a single oil, itself a non-random mixture of triacylglycerols, or on a blend of oils. An old example of the first of these is the interesterification (randomisation) of lard. This fat is unusual in having a high level of palmitic acid in the sn-2 position. During reaction with sodium methoxide the lard fatty acids become randomly distributed. Overall there is no change in fatty acid composition but a change in triacylglycerol composition. As a consequence of this modification the randomised lard is a better shortening. More commonly, interesterification is applied to a blend of oils. The oils that are mixed can differ in their level of unsaturation (an unsaturated oil and a hydrogenated oil), in their range of chain length (such as a lauric oil and a $C_{16/18}$ oil), or in the presence of a less common fatty acid such as γ-linolenic acid, EPA, or DHA in one of the components.

Mixtures of esters may be interesterified with an appropriate basic catalyst. When this happens the natural non-random mixture is ultimately converted to a randomised mixture. The change will be even more marked with a mixture of two different types of oils such as a lauric oil and a non-lauric oil. Alteration is particularly apparent in the fatty acids present in the sn-2 position. Before interesterification of a vegetable oil these will be mainly unsaturated acids but after complete randomisation the fatty acids at all three positions will be the same. These changes have important effects on the physical (particularly melting behaviour) and nutritional properties of the modified fat.

Interesterification is used in newer methods of producing spreads with a reduced content of *trans* acids. Hydrogenation producing large amounts of *trans* acids, can be replaced by interesterification of a soft fat with a hard fat. Most spreads produced in Europe now have a very low level of *trans* acid though this may be less true for cooking fats and industrial spreads. When randomisation is complete triacylglycerol composition can be calculated from the fatty acid composition since the amount of an individual triacylglycerol (ABC) will depend only on the proportions of each of these acids (a%, b%, c%, respectively). The level of this single glycerol ester will be 100 [a/100 $\times$ b/100 $\times$ c/100]% and the level of all isomers having one A, one B, and one C acyl chain will be six times this figure since there are six stereoisomers meeting this requirement.

Ester–ester interchange can be achieved without a catalyst at temperatures above 200°C but is usually carried out at 20–100°C with a basic catalyst such as sodium hydroxide, sodium methoxide, or sodium potassium alloys. At ~80°C the reaction takes 30–60 min and may be carried out on a multi-tonne scale. The oil should be free of water, carboxylic acid, and hydroperoxide as these compounds will destroy the catalyst. The true catalyst is thought to be either a diacylglycerol anion [ROCOCH$_2$CH(OCOR)CH$_2$O$^-$] formed by interaction of a triacylglycerol molecule with sodium methoxide or the enolate [ROCOCH$_2$CH(OCOR)CH$_2$OCO$^-$=CHR′] resulting from removal of a proton from the α-methylene function. The product will contain some free acid if NaOH is used as catalyst or methyl ester, if the catalyst is NaOMe and must be refined to remove these. The amount of catalyst must be as low as possible to minimise these losses.

Directed interesterification is a modification of the normal process when the reaction is conducted at a lower temperature (25–35°C). Under these conditions the less soluble triacylglycerols crystallise from the solution. This disturbs the equilibrium in the liquid phase and this will be continually re-established. The consequence is to raise the levels of SSS and of UUU triacylglycerols in the final interesterified product.

With enzymatic catalysts, interesterification leads to structured lipids of the types described below.

Nutritional and other physiological properties of fats (triacylglycerols) depend on their detailed structure. Until now, use has been made of materials provided by agriculture modified in minor ways such as fractionation, partial hydrogenation, and interesterification as described in earlier sections of this chapter. Individual triacylglycerols can be synthesised in the laboratory in modest quantities but these will not usually be appropriate for large-scale human consumption. By exploiting the specificity of lipases it is now possible to produce large quantities of oils and fats approximating to a specification designed to optimise some important physical and/or nutritional property.

The cost of the enzyme is generally so high that their use is only economic for high-value products but these difficulties are being overcome as enzyme producers develop immobilised enzymes of greater stability with a longer useful life. At the same time our understanding of enzyme structure allows changes to be made leading to enhanced selectivity. Beyond this there is growing willingness to pay more for fats for which approved health claims can be made. Enzymatic reactions are considered to be more 'natural' or 'greener'.

Lipases show several different kinds of specificity which can be exploited. The most common is 1,3-regiospecificity in which reaction is confined to the sn-1 and 3 positions of a triacylglycerol with no change at the sn-2 position. Lipases may also show specificity for selected fatty acids with which they associate. This specificity may depend on the unsaturated centres and especially the position of the double bond closest to the carboxyl group or may be related to chain length.

There are many reports of preparations of structured triacylglycerols which can be achieved in one-step or two-step processes. In the former, an oil or fat has some of its sn-1/3 acyl chains replaced by different fatty acids. The acyl donor may be an ester such as an alkyl ester or triacylglycerol mixture or an acid (acidolysis). The reaction is illustrated in very simple form in the following equation. In reality the reaction is more complicated since neither reactant will be the individual species indicated in the equation.

Gl-ABC + D (as acid or ester) $\rightleftharpoons$ Gl-DBC + Gl-ABD + Gl-DBD

The products represent racemic triacylglycerols with acyl chains A–D. The sn-2 position is unaffected by this process.

Good results have been obtained with lipases such as those from *Rhizomucor miehei* (Lipozyme), *Rhizopus delemar*, and *Candida antarctica*. Incorporation of reactant is usually in the range 40–65%. This procedure is simpler than the two-step reaction but the products are less pure.

In the two-step process triacylglycerols are selectively deacylated at the *sn*-1 and 3 positions by enzyme-catalysed ethanolysis and pure 2-monoacylglycerol is isolated. The monoacylglycerol is then acylated at the free hydroxyl positions using a 1,3 specific lipase and an appropriate acyl donor which may be free acid, alkyl ester, or vinyl ester.

$$Gl\text{-}ABC \rightarrow Gl\text{-}(OH)B(OH) \rightarrow Gl\text{-}DBD$$

Two-step synthesis of a structured triacylglycerol proceeding through a 2-monoacylglycerol which is isolated and purified before the second step. Gl stands for the glycerol residue; OH represents the free OH groups in a monoacylglycerol; and A, B, C, and D are acyl groups.

There is a greater control of the reaction when this is conducted in a carefully selected solvent but this involves additional handling and cost and the aim is to produce bulk products having the desired properties as simply and cheaply as possible.

Typically a plug-in reactor ($1\,m^3$), containing Lipozyme TL IM prepared from *Thermomyces lanuginose* lipase is supplied by the enzyme producer. Two oils such as palm oil (or palm stearin) and palm kernel (or coconut) oil are passed through the reactor and emerge 1 h later as interesterified oil. The reaction occurs at 70°C which is 30°C lower than for chemical interesterification, no downstream processing is required, and the product has no acids with *trans* unsaturation. This approach is being used to produce spreads with a low level of *trans* acids.

Many laboratory experiments are concerned with attempts to produce triacylglycerols of the type MLM where M is an easily metabolised fatty acid of medium-chain length (frequently C_8) and L is a long-chain acid including nutritionally important fatty acids such as EPA or DHA.

The nature of the fatty acid in the *sn*-2 position is controlled by the selection of starting material. Vegetable oils will provide sources of oleic and linoleic acid in this position and fish oils will be used for the long-chain polyunsaturated fatty acids. In a preliminary stage the levels of these important acids may be enhanced prior to inter-esterification (Section 2.7).

Human milk fat is unusual in that it is rich in triacylglycerols containing a saturated acid (palmitic) in the sn-2 position. This is unusual among natural fats so a product with this structural fea-ture called 'Betapol' has been developed for addition to infant for-mula. In theory, tripalmitin is reacted with unsaturated acids in the presence of a 1,3-stereospecific lipase (from *Rhizomucor miehei*). In practice the reactants are a palm stearin rich in tripalmitin and a mixture of canola and sunflower oils rich in oleic acid.

Nutritional Properties

7.1   Introduction

About 153 million tonnes of 17 commodity oils and fats were produced in 2006/07. It has generally been assumed that around 80% is used for human food (122 million tonnes) but the increasing demand for biodiesel has probably reduced this to a figure close to 115 million tonnes. However, this figure is too high as a measure of fat consumption by reason of loss and waste but also too low because of other fat sources not included in statistical tables for the commodity oils such as cheese and meat. It is not easy to derive a world figure for fat consumed by humans.

Many references in the media suggest that fats are dangerous compounds which should be removed from the diet altogether. *This is completely false.* As with other dietary components, there are appropriate levels of intake and other levels associated with ill health because intakes are too high or too low. Fats and oils are an essential part of the human diet for the following reasons:

- They are the most efficient source of energy.
- They are a source of many bio-active compounds and they contain important acids (essential fatty acids, EFA, see Section 7.2) which animals need but cannot bio-synthesise and must ingest from plant sources.
- They are carriers of important minor components such as fat-soluble vitamins and phytostcrols.
- Fats add palatability to our food and contribute flavour and texture.

The body of a lean man of 70 kg is made up of water ($\sim$42 kg, 60%), protein and fat (each $\sim$12 kg, 7%), with a balance (including minerals) of ~4 kg (6%). In an obese man of 100 kg, 35 kg will be fat. Put differently, of the extra 30 kg, 23 kg is additional fat. A woman's body has less water and more fat (~27%).

Optimum energy intakes have to be related to life style. Some among the very poor have too few calories while many in developed countries have too many for their present low-energy lifestyle, leading to overweight and obesity. When energy is needed it comes first from the limited supplies of glycogen in the body (~1.5 kg), then from stored fat, and ultimately, in cases of severe starvation, from protein. These three major nutrients have average energy levels as indicated: fat (38 kJ/g or 9 kcal/g), carbohydrate (17 kJ/g or 4 kcal/g), and protein (16 kJ/g or 3.8 kcal/g). Clearly, fat provides twice the energy of the other two major nutrients.

Dietary fat supplies come from obvious sources such as butter and spreads and salad oils and less obviously from other dairy products such as cream and cheese, meat and fish, baked goods that contain fat (cakes, biscuits, pastries, breads), chocolate and other confectionery items, and fried foods where frying oil has been added to the fat in the food itself. It is not easy to know how much fat is being consumed but weight and girth can be measured by reference to scales or to waistline measurement. In western countries it is recommended that fat consumption should not exceed 30% of energy requirements but for many persons intake is closer to 35% or even 40%. For a food intake corresponding to 2000 kcal/day fat intakes of 30%, 35%, and 40% require 67, 78, and 89 g/day of fat. In rice-consuming countries dietary customs are different and fat consumption is generally lower. Table 7.1 provides data on fatty acid composition of a small selection of foods. Further information can be gathered from the references provided with the Table.

Table 7.2 provides figures on 'disappearance' in units of kg/person/year over a 20-year period for the main consuming countries/regions. As explained in the footnote to this table disappearance is not the same as dietary consumption. For long, human consumption of oils and fats was believed to average 80% of total usage. With the growing industrial use of these commodities for biodiesel the global average is falling below 80% and is probably now just below 75%. Also this average varies between countries and depends mainly on the size of their oleochemical industry. The traditional oleochemical industry, producing mainly soap and other surface active compounds for personal care and for cleaning, is concentrated in USA, Western

Table 7.1 Levels of total fat and the major types of fatty acids (cited as g/100g of food) in selected foods taken from the publication Gunstone (2004) itself adapted from the original and larger listing of McCance and Widdowson (1998)

Food	Fat content	Saturated	cis-Monoun-saturated	Poly omega-6	Poly omega-3	Trans Total
White bread	1.9	0.40	0.25	0.62	0.04	0.00
Croissants	26.0	14.33	6.62	1.00	0.41	1.64
Danish pastry	14.1	8.57	1.54	1.65	0.24	0.84
Whole milk	4.0	2.48	0.93	0.10	0.02	0.14
Semi-skimmed	1.7	1.07	0.39	0.05	0.01	0.07
Double cream	53.7	33.39	12.33	1.34	0.48	1.83
Cheddar cheese	32.7	19.25	7.14	0.99	0.28	3.10
Cooking fat	99.5	24.52	31.56	27.06	1.49	10.37
Butter	82.2	52.09	18.48	1.41	0.68	2.87
Spread (70%)	70.0	9.44	31.63	11.73	3.79	10.45
Soybean oil	99.9	15.60	21.20	52.50	7.30	Trace
Olive oil	99.9	14.3	73.00	7.50	0.70	Trace
Lean beef cooked	8.2	3.26	3.41	0.36	0.09	0.28
Chicken roasted	3.7	1.02	1.58	0.60	0.13	0.05
Sausage grilled	19.5	7.69	8.35	1.26	0.15	0.39
Cod raw	0.7	0.13	0.08	0.02	0.26	0
Potato chips	11.0	5.96	2.69	0.16	0.01	0.43
Potato crisps	34.2	14.04	13.51	_a	_a	_a
Peanuts	46.0	8.66	22.03	12.75	0.35	0
Salad cream	31.0	3.29	11.44	13.55	1.00	0.10

[a] Present but not in known amount.

These figures originate in Supplement to McCance and Widdowson's 'The Composition of Foods', Ministry of Agriculture, Fisheries, and Food and The Royal Society of Chemistry, 1998, and are based on 550 results obtained between 1990 and 1997. The analyses are mean values measured on several samples. Full details of fatty acid composition are given in the book and some foods are detailed in the raw and cooked state. Some of the figures for prepared foods will be different from what they were in the 1990s.

Similar information is available from American sources (Hands, 1996 in *Bailey's* and USDA nutrient database <www.nal.usda.gov/fnic/foodcomp>).

Europe, and Japan to which must now be added the oil palm-growing countries of South East Asia, particularly Malaysia. The proportion of available oils and fats used for human consumption will be greater than 75% in countries with little or no oleochemical industry. The table shows how *per capita* consumption has increased over time and how it varies between developed and developing regions of the world. Total demand for dietary fat (and for meat) increases with population and with income and is thus expected to increase

Table 7.2 Changing average 'disappearance' (kg/person/year) over 20 years in the four largest consuming countries/regions

Country	1986	1996	2006/07
World	14.4	16.9	23.0
USA	38.5	45.2	54.3
EU	36.4	43.3	56.7
China	6.3	11.2	21.7
India	7.4	9.3	11.9

Note: The term 'disappearance' relates to the total quantity of 17 commodity oils used for both food and non-food purposes.
Figures for EU relate to EU-15 in 1986 and 1996 and to EU-27 in 2006/07.
Source: *The Revised Oil World 2020 – Supply, Demand and Prices* (2002) and *Oil World Annual 2007* (2007) both published by ISTA Mielke GmbH, Hamburg, Germany.

for many years to come. It is of interest that developed countries like Australia and New Zealand with limited oleochemical industry had disappearance levels of 37.7 and 36.9 kg/person/year, respectively in 2006/07. These values must be close to annual dietary fat consumption in those countries though as already indicated they are too high as a measure of consumption by reason of loss and waste but also too low because of other fat sources not included in statistical tables.

It is not only total fat intake that needs to be controlled but also the nature of that fat, usually expressed in proportions of the type of fatty acids present. Concepts have moved over time partly as a consequence of changing availability of dietary fats such as the shift from animal to vegetable sources, partly as a result of a better understanding of the link between diet and health/disease, and particularly because of concern over cardiovascular disease (CVD). There is a range of risk factors for CVD that include, but is not confined to, dietary fat (Section 7.13).

7.2   EFA and fatty acid metabolism

The fatty acids present in our bodies as triacylglycerols and other lipid classes come from endogenous or exogenous sources. Either we make them in the body (endogenous) from precursors such as acetate (resulting from the catabolism of fats or carbohydrates) or we get them directly from the fats in our diet (exogenous). In addition, the body is able to modify the structure of fatty acids mainly by chain elongation (adding two carbon atoms through a metabolic cycle) or by desaturation (inserting double bonds). However, there are some acids that humans, in common with other animals, cannot make and these must be dietary in origin and come from plant sources (or from eating the flesh of animals that have already consumed these acids). These are described as EFA. Once ingested, EFA may be metabolised to other physiologically important acids. Since these processes are not always efficient there is debate as to whether the important metabolites themselves should also be considered as EFA.

The major biosynthetic pathways to fatty acids involve three stages:

(1) *De novo* synthesis of palmitic (or other alkanoic) acid from acetate (C_2, a product of carbohydrate metabolism) by reaction with malonate (C_3), itself formed from acetate.
(2) Further chain elongation of saturated or unsaturated acids by one or more two-carbon units.
(3) Desaturation: particularly of stearic acid, first to oleic acid (18:1), and then to linoleic (18:2) and linolenic (18:3) acids.

Further sequences of elongation and desaturation producing discrete families of polyunsaturated fatty acids.

Whether in plants or animals these changes take place in different parts of the cell, under the influence of specific enzymes or enzyme complexes, and require the acids to be in appropriate substrate form.

While animals can produce their own supplies of fats they get these mainly as part of their dietary intake. Much of this exogenous fat will be metabolised through oxidation to produce energy, stored in part as ATP (adenosine triphosphate), but some will be stored as lipid, perhaps after modification. Phospholipid bilayers are fundamental components of all living matter and these must contain particular fatty acids.

7.3 *De novo* synthesis of saturated acids

In plant systems *de novo* synthesis occurs in the plastid and results mainly in the conversion of acetate to palmitate. All 16 carbon atoms in palmitic acid are derived from acetate. In this pathway acetate (CH_3COOH) and malonate ($HOOCCH_2COOH$) react through a series of steps converting acetate first to butanoate (C_4), then to hexanoate (C_6), and sequentially thereafter, two carbon atoms at a time, to palmitate (C_{16}). At this stage a thioesterase liberates the acyl chain from ACP. The thioesterase is not completely chain-length specific and acids of other chain lengths may be produced. This is obviously true in the lauric oils where the major saturated acid (lauric, 12:0) is accompanied by lower levels of caprylic (8:0), capric (10:0), myristic (14:0), and palmitic acid (16:0).

7.4 Desaturation and elongation in plant systems

The first desaturation of a saturated acyl chain occurs in the plastid. The most common is the conversion of stearate (18:0) to oleate (18:1) and involves the removal of hydrogen atoms from C-9 and C-10 to give a *cis* olefinic bond under the influence of a Δ9 desaturase. The system is oxygen-dependent and involves the reduced form of ferredoxin.

Other saturated acids can be desaturated similarly so there is a group of Δ9 monoene acids such as myristoleic (9c-14:1), palmitoleic (9c-16:1), oleic (9c-18:1), and gadoleic (9c-20:1) with each unsaturated acid made by Δ9 desaturation of the corresponding saturated acid.

Elongation by two carbon atoms occurs commonly in fatty acid biosynthesis. It is a variant of *de novo* chain lengthening and occurs with acetyl- or malonyl-CoA or their ACP derivatives. The substrate is any pre-formed saturated or unsaturated acid. For example, erucic (22:1) in high-erucic acid rapeseed oil and nervonic acid (24:1) in honesty seed oil are formed from oleic acid by two and three elongations, respectively. These belong to a family of omega-9 monoene acids.

$$18:1\ (9) \rightarrow 20:1\ (11) \rightarrow 22:1\ (13) \rightarrow 24:1\ (15)$$
$$\text{oleic} \qquad \text{cetoleic} \qquad \text{erucic} \qquad \text{nervonic}$$

Further desaturation in the cytoplasm converts oleate (18:1 in the form of a phosphatidylcholine) to linoleate (18:2) with a Δ12 desaturase and converts linoleate (as its monogalactosyldiacylglycerol derivative) to linolenate (18:3) with a Δ15 desaturase. *These desaturation steps are confined to plants and do not occur in animals.* They are very important changes as linoleate and linolenate are precursors of important acids with significant physiological properties. The additional double bonds have *cis* configuration and are in a methylene-interrupted relation to each other. This 1,4 diene unit is characteristic of polyunsaturated fatty acids and is to be distinguished from the 1,3 (conjugated) systems in carotenoids and the 1,5 system in terpenes.

$$-CH=CH(CH_2)_nCH=CH-$$

$n = 0$ (conjugated 1,3), 1 (methylene-interrupted 1,4), or 2 (bismethylene-interruptrd 1,5)

Though common in animal systems $\Delta 6$ desaturase is less apparent in the plant world though not completely absent. It operates in the biosynthesis of γ-linolenic acid (6,9,12-18:3) from linoleate and of stearidonic acid (6,9,12,15-18:4) from α-linolenate. The C_{20} and C_{22} polyenes that characterise animal systems and particularly fish lipids either do not exist in plant systems or are exceedingly rare. The production of important acids such as arachidonic (5,8,11,14–20:4), eicosapentaenoic (5,8,11,14,17–20:5), and docosahexaenoic (4,7,10,13,16,19–22:6) in plant systems is a challenge for plant geneticists. Research has got as far as proof of concept but much remains to be done before an economically viable system is produced.

7.5 Desaturation and elongation in animal systems

Families of polyene acids are produced by a combination of elongation and desaturation processes starting with palmitoleic acid (omega-7 family), oleic (omega-9 family), linoleic (omega-6 family), and linolenic acid (omega-3 family). The acids in each family share a common structural feature viz. the position of the double bond closest to the methyl end of the molecule. These changes are particularly important in animal systems and lead to the long-chain polyunsaturated fatty acids that are of considerable nutritional significance. The changes occurring in mammalian systems are set out in Figure 7.1. The same enzymes are used in each family and there is competition for access to these. The ratio of omega-6 to omega-3 acids required in the diet for optimum health is a matter of present debate (Section 7.10).

Omega-6 family		Omega-3 family	
18:2 (9,12)	linoleic	18:3 (9,12,15)	α-linolenic
↓ Δ 6-desaturase		↓ Δ 6-desaturase	
18:3 (6,9,12)	γ-linolenic	18:4 (6,9,12,15)	stearidonic
↓ elongation		↓ elongation	
20:3 (8,11,14)		20:4 (8,11,14,17)	
↓ Δ 5-desaturase		↓ Δ 5-desaturase	
20:4 (5,8,11,14)	arachidonic	20:5 (5,8,11,14,17)	eicosapentaenoic
		↓ elongation	
		22:5 (7,10,13,16,19)	
		↓ elongation	
		24:5 (9,12,15,18,21)	
		↓ Δ 6-desaturase	
		24:6 (6,9,12,15,18,21)	
		↓ β-oxidation	
		22:6 (4,7,10,13,16,19)	docosahexaenoic

Figure 7.1 The omega-6 and omega-3 families of polyunsaturated fatty acids.

The most significant acids in these sequences are linoleic and arachidonic in the omega-6 family and α-linolenic, eicosapentaenoic (EPA), and docosahexaenoic acids (DHA) in the omega-3 family, The two C_{20} acids are precursors of an important group of eicosanoids including the prostaglandins and leukotrienes. The numbers in parenthesis indicate the positions of the double bonds all of which have the *cis* configuration.

7.6   Antioxidants (see also Section 6.2)

It is widely accepted that for the most part lipid oxidation occurring via reactive free radicals is an undesirable process. In fat-containing foods oxidation leads to unacceptable flavours and the foods are ultimately rejected. *In vivo*, radicals react with proteins and nucleic acids. The consequent changes are linked to several disease conditions and dietary antioxidants are considered to be important. It is for this reason that the presence of antioxidants of different kinds in several dietary components is loudly acclaimed.

It is generally assumed that the higher the concentration of anti-oxidant the greater the value of that product. But the link between particular antioxidants and good health or disease has not been clearly proved and we are some way from knowing the optimum levels required for human health. Results in the laboratory do not always transfer simply to the more complex reality of human life.

7.7   Cholesterol and phytosterols

Sterols are important minor components of most oils and fats (Section 1.6.). Sources of plant origin contain a range of phytoster-ols while those coming from animals are rich in cholesterol. Typical levels of the latter for lard (0.4%), beef fat (0.1%), mutton tallow (0.2–0.3%), and butter (0.2–0.4%) are indicated in parenthesis. Eggs contain ~300 mg of cholesterol per egg. The sterol may be present as free sterol or associated with a fatty acid as sterol ester.

The human body contains ~100 g of cholesterol and requires about 1 g of new cholesterol each day. Total cholesterol represents ~0.2% of body weight with one-third in the brain and nervous tissue, one-third in muscular tissue, and the remainder in cell membranes. The daily requirement will be mainly of endogenous origin (600–1000 mg) with the balance from dietary sources (250–500 mg) coming mainly from eggs (~300 mg per egg) and animal fats. Only about one half is absorbed. Phytosterols interfere with the absorption of cholesterol and are being added to spreads to reduce cholesterol uptake.

Cholesterol is an important and necessary compound used *in vivo* for the production of bile salts (emulsifiers) in the liver, steroid hormones (*e.g.* sex hormones) in the adrenal glands, and of vita-min D in the skin.

Elevated cholesterol levels in blood are recognised as one of several risk factors in CVD. This level can be easily measured and is frequently used as an index of cardiovascular health. Levels vary between individuals. They are only slightly influenced by dietary intake of cholesterol and rather more by saturated fatty acids (SFA) and by non-dietary influences. Levels above ~230 mg/100 ml of blood (~6.0 mmol/l) are considered to be undesirable in terms of CVD (Section 7.13) but levels below 160–180 mg/100 ml may lead to risk of non-cardiovascular death.

The possible link between cholesterol oxidation products and coronary heart disease (CHD) and other disease states makes it appropriate to discuss the source and formation of such compounds. Cholesterol (Section 1.5) contains a cyclic double bond ($\Delta5$) and two tertiary carbon atoms in its side chain (C-20 and C-25), all sites where oxidation may occur. Some cholesterol oxides are produced as part of the normal metabolism of cholesterol to bile acids but at higher levels they affect human health by contributing to the development of atherosclerosis. When cholesterol oxides replace cholesterol in the cell membrane they alter its fluidity, permeability, stability, and other properties. Oxidised animal-based foods represent a primary source of oxidised cholesterol. Such products are not present in fresh foods but are formed during handling prior to consumption, mainly through autoxidation (Section 6.2).

7.8 Conjugated linoleic acid

C_{18} acids with two double bonds conjugated with each other were recognised as trace components of milk fat over 50 years ago and have engendered renewed interest in recent years following the recognition of their potential value in the treatment of cancer, obesity, and diabetes.

Conjugated linoleic acid (CLA) has been identified at low levels in milk fat (3–6 mg/g of total fat), butterfat (12–14 mg/g), cheeses (2–20 mg/g), and in lamb and beef meat (4–5 mg/g). Several isomers may be present. The major component (the 9c11t isomer) is designated rumenic acid because of its formation in the rumen of the cow. It is believed to be a metabolic product resulting from linoleic acid by two linked pathways: isomerisation of linoleic acid (9c12c-18:2) and $\Delta9$-desaturation of vaccenic acid (11t-18:1). The 7t9c and 10t12c dienes are also present at lower levels along with many other isomers.

$$9c\ 12c\text{-}18{:}2 \rightarrow 9c11t\text{-}18{:}2 \rightarrow 11t\text{-}18{:}1 \rightarrow 9c\ 11t\text{-}18{:}2$$

linoleic acid   rumenic acid   vaccenic acid   rumenic acid

CLA can be made in larger volumes and higher concentrations by alkali isomerisation of linoleic-rich vegetable oils such as safflower. The product contains two isomers (9c11t and 10t12c-18:2) as virtually the only CLA present along with unreacted palmitic, stearic, and oleic acids from the starting material. These two CLA isomers show different physiological properties. The products of the isomerisation process are free acids and these are generally converted to triacyl-glycerols before being used in human or animal diets. This can be done enzymatically with lipases such as those from *Mucor miehei* or *Candida antarctica.* Esterification then proceeds under mild conditions without modification of the double-bond systems in the CLA.

Several potential benefits have been claimed for CLA. Dietary supplementation has been shown to reduce the number and size of mammary tumours with the 9c11t compound probably the more effective isomer. CLA has also been used to alter body composition since it is claimed to reduce body fat, increase lean mass, and improve feed efficiency. These effects are associated particularly with the 10t12c isomer. Positive results have been obtained with young animals and this is of interest to those concerned with meat production. Conclusions are less certain with humans.

It has also been claimed that in animals CLA affects the immune function and has an effect on bone remodelling. In animal husbandry attempts have been made to increase CLA levels and also to decrease fat production mainly by adjustment of dietary regimen. Increasing CLA levels in poultry meat and in eggs is seen as a potential method of increasing levels of CLA in the human diet.

7.9 Diacyglycerols

A cooking oil rich in diacylglycerols has been available in Japan since 1999. It is made by reaction of glycerol or 1-monoacylglycerol with fatty acids from natural edible oils using a 1,3-regiospecific lipase. The product is at least 80% diacylglycerol of which ~70% is the 1,3-isomer. This figure should be higher but some acyl migration occurs during the refining processes required to obtain the final product. Similar products are now available in the USA and elsewhere.

Acylation of glycerol to produce 1,3-diacylglycerols.

There is also evidence that phytosterols may be more effective in reducing blood cholesterol levels when taken in diacylglycerols rather than in triacylglycerols because of their greater solubility in the former. A diacylglycerol-rich oil with added phytosterols, is sold with permitted claims such as 'it is less likely to become body fat' and 'it lowers blood cholesterol levels'.

The differing physiological effects of di- and triacylglycerols, observed in both animals and humans, are not caused by differences in energy values (38.9 kJ/g and 39.6 kJ/g, respectively) or of absorption of the fates (both 96.3%) but by their different metabolic fats after absorption. 2-Monoacylglycerols, resulting from normal triacylglycerol digestion, are readily reconverted to triacylglycerols in epithelial cells but the 1-monoacylglycerol resulting from the diacylglycerol prepar-ations are only poorly re-esterified. In animal studies the structural differences between di- and tri-acylglycerols markedly affect their nutritional properties including body fat accumulation, serum lipid profile, and the development of hyperinsulinemia and hyperleptinemia.

Double-blind clinical studies in humans have shown that it is possible to reduce the magnitude of postprandial lipemia by consuming diacylglycerols in place of triacylglycerols and therefore the former may be less atherogenic than the latter. Dietary diacylglycerols may therefore reduce the risk of coronary arteriosclerotic diseases by lessening the postprandial increase in the concentration of remnant-like lipoprotein particles.

In contrast to triacylglycerols, diacylglycerols suppress body weight and regional fat deposition, both visceral and hepatic. In a study of obese Americans, decreases in body weight were significantly higher in patients consuming diacylglycerols compared to those on triacylglycerols. It has also been suggested that the consumption of diacylglycerol-rich oil might be useful for maintaining the quality of life of patients with diabetes.

7.10 Recommended intake of fats and of fatty acids

Dietary requirements apply to healthy adults. Different recommendations may be appropriate for those who are not in good health, for babies and children under 5 years, for pregnant and lactating females, and for older people. Dietary advice to the general population should always be part of a package that includes advice to stop smoking, to take more exercise, to maintain a healthy weight, and to relieve stress. Frequently there is undue emphasis on fat consumption and the other risk factors are overlooked.

Dietary recommendations have been made in most developed countries. These tend to be fairly similar. This is not surprising for it takes a measure of courage (or stupidity) to differ from everyone else. It seems likely that few of those giving the advice have read the original research reports on which the advice is based. Many of these recommendations are based on the need to lower plasma cholesterol levels to reduce CHD. The following statements seek to present an overview of the general recommendations relating to fat. In summary they represent a call to limit the consumption of total fat and of saturated fat and to improve the omega-6/omega-3 ratio (when we can agree on what that should be).

Although humans have the ability to synthesise most (but not all) fatty acids, most of our fat is dietary in origin. Some of this is metabolised to produce energy, some is laid down as adipose tissue, and some finds its way into cell membranes as phospholipids. The fatty acids may be modified before being laid down.

Total fat intake ranges from 15% to 40% energy across most of the world's population. Some have argued that the maximum should be 35% as a practicable first step for those consuming in excess of this figure. The prevalent view is that the maximum should be 30 energy% and that it may be desirable to be below this level.

Prior to World War II most people in developed countries depended mainly on animal fats such as butter, lard, and tallow but since around 1950 this has changed so that present day diets are dominated by fats of plant origin. Nevertheless many still consume a significant portion of their dietary fat from dairy products (milk, cream, butter, cheese) and from meat.

Concern about the growing incidence of CHD in developed countries led to changes as to what was considered to be a healthy fat intake. The first fatty acids to be targeted were the saturated acids. It had been known since the mid-1950s that there is a link between saturated acids, blood cholesterol levels, and CHD and that replacement of dietary saturated acids by mono- or poly-unsaturated acids led to a reduction of cholesterol level and of the incidence of CHD. This accelerated the changeover from animal to vegetable fats and the development of soft spreads containing less saturated fat and more polyunsaturated fat as an alternative to butter.

It is now generally agreed that saturated acids should not exceed one-third of total fat intake giving a figure of 10% in a total of 30% though there are those who argue for lower levels (around 8%). However, this is too simple as all saturated acids do not have the same effect on cholesterol level. Acids with up to 10 carbon atoms in their molecules (present in dairy fats and in lauric oils) are metabolised in a different way from other saturated acids and are not cholesterol-raising. Also, stearic acid is considered to have only a small effect, possibly because it is so easily metabolised to oleic acid through the activity of the ubiquitous $\Delta 9$-desaturase. This leaves lauric (12:0), myristic (14:0), and palmitic (16:0) acids. Myristic acid is considered by some to have the greatest cholesterol-raising influence but this is generally a minor dietary component compared with the more common palmitic acid. One factor that has not yet been clearly defined is how far it matters whether these acids are present in the sn-2 or sn-1/3 positions of triacylglycerol molecules. This factor is not likely to be taken seriously so long as saturated acids are listed as a single class on food labels.

Monounsaturated acids with cis configuration are believed to be the safest of all fatty acids. This refers to oleic acid and to the smaller amounts of dietary palmitoleic acid (16:1). Longer-chain acids such as erucic acid (22:1) are considered to be less desirable. This acid was the major fatty acid in traditional rapeseed oil (now called high-erucic rapeseed oil, HEAR) but this has been bred out of canola and other rapeseed oils used for food purposes. HEAR is still produced but mainly for industrial purposes.

In recent years it has been accepted that monounsaturated acids with trans unsaturation are undesirable dietary components because of their effect on cholesterol levels. They raise LDL (low-density lipoprotein) levels and lower HDL (high-density lipoprotein) levels and so change the important ratio of these two in an undesirable direction. There are two sources of dietary trans acids: dairy (ruminant) fats and partially hydrogenated vegetable oils. Both contain a range of trans 18:1 acids but differ in detailed composition. The presence of vaccenic acid (11t-18:1) in dairy fats may be of less concern because it is readily converted to 9c11t-18:2 (rumenic acid, a CLA) which has favourable attributes. The concern about trans acids led to labelling requirements in the USA in 2006 with the consequential reformulation of many fat-containing foods to minimise levels of trans acids or to have them at a level that allows a claim for zero trans (<0.5 g per serving of 14 g).

Sometimes this lowering of *trans* is accompanied by a small rise in the level of saturated acids in order to achieve the necessary melting behaviour.

For polyunsaturated fatty acids early recommendations for all acids of this class taken together have been replaced by separate recommendations for omega-6 (linoleic acid and its long-chain metabolites) and omega-3 acids (linolenic acid and its long-chain metabolites). It is now being asked whether there should even be separate targets for C_{18} members and for C_{20} and C_{22} members because, particularly in the omega-3 series, there is concern over the limited ability of animals to convert ALA (itself frequently consumed at less than optimum levels) to the very important EPA and DHA. This is more apparent in males than in females. Particularly because of the high availability of linoleic-rich soybean oil and sunflower oil and the competition that exists between metabolism of linoleic and linolenic acids, there is concern that present omega-6/omega-3 ratios are too high. This is a problem in those countries such as the USA where there is a high consumption of linoleic acid (via soybean oil) and a low consumption of fish (see Chapters in Breivik, 2007).

Dietary intake of omega-6 PUFA will be mainly as linoleic acid and only to a small extent as arachidonic acid (AA) (20:4). It has been suggested that the omega-6 acids should be around 6 energy% with 3–10% representing a safe range but there is a growing opinion that even these levels might be too high. Arguments have been put forward for a more limited intake between 2% and 3%. It has been calculated that with about 20–40% of human body mass being fat and 15–20% of body fat being linoleic acid an adult American will have more than 3 kg of linoleic acid in his tissues. This large reservoir of omega-6 acid may need 3 years to equilibrate with a change in dietary lipids.

For omega-3 acids separate recommendations are now available for linolenic acid and for EPA/DHA. One authority suggests a total intake of 1 energy% (within a range of 0.5–2.5%) for linolenic acid and 0.5 % (within a range of 0–2.0%) for EPA and DHA. Another group have expressed this slightly differently at 1.0% for linolenic acid and 0.3% for the higher acids with a minimum of 0.1% for each of EPA and DHA. A further recommendation is for a minimum daily intake of 450 mg of EPA and DHA. Many food labels indicate levels of omega-3 acids but do not distinguish between 18 and 20/22 acids. This generally means that linolenic acid is the only omega-3 acid present and may confuse the consumer.

The use of the omega-6/omega-3 ratio has been questioned. It is now considered that with the present ratio of between 5 and 10 it is better to be closer to the lower end of this range. Some have argued for ratio levels below 5.

The balance of dietary fat should consist of *cis*-monounsaturated acids (almost entirely oleic acid) and this becomes particularly important when total fat intake exceeds 30%. It is considered that oleic acid should represent 11–16% of dietary intake but should not exceed this upper limit.

Because DHA is such an important component of the brain appropriate supplies of this acid are required by a developing foetus and are supplied by the mother via the placental cord. This need continues after birth and should be met through human breast milk that, in a well-nourished mother, will contain adequate levels of DHA. Problems arise when the child is not breast-fed and, more significantly, in pre-term infants since brain development is particularly marked in the final trimester before birth. It is now recommended that during pregnancy and lactation dietary fat intake should be as for other adults except that the intake of DHA should be at a minimum level of 300 mg/day. Infant formula for Western countries, expressed as percentage of fatty acids, should contain linoleic (10.0%), α-linolenic (1.50%), arachidonic (0.50%), DHA (0.35%), and EPA (<0.10%). Slightly different levels are used in Japan with linoleic acid at 6–10% and DHA at 0.6%. The American Heart Association now recommends that patients with known CHD consume ~1 g/day of EPA and DHA. For those without known CHD the recommendation is for at least two servings of fish (preferably oily fish) each week – equivalent to 500 mg/day. Other recommendations range from 200 to 650 mg/day for the general population, with higher levels up to 1000 mg/day for those at risk of CVD. A recent UK proposal is for 450–1800 mg/day.

7.11   Role of fats in health and disease

Modern medicine and modern standards of hygiene have freed the world, or at least many parts of it, from the killer diseases of previous centuries such as tuberculosis, smallpox, and diphtheria and we inhabit a world where increasing numbers live healthier and longer lives. This has highlighted the diseases that remain and are killers, not only of the old, but of men and women in their prime, particularly CHD and cancer. We are increasingly aware that many diseases that remain, whether killers or not, are related in some part to lifestyle of which diet, pollution of the environment, and level of physical activity are all important factors.

The following sections are related to the role of fat in some disease conditions. However, it is important to realise that fat is only part of our diet and that diet is only part of the problem and of the solution. It is inadvisable to focus on a single issue and ignore others. Fat has a very negative image at the present time and we need to correct that. Fat is an essential part of the diet and is linked to good health as well as to disease. It is important to optimise the quality and the quantity of fat consumed in relation to other aspects of lifestyle. This implies: that we know what fats we should consume and in what quantity, that agriculture and the food industry can supply these, and finally that we can persuade human populations to choose healthy and affordable diets. We must recognise that we are a long way from discovering the final truth in respect of dietary fat and that what is written in the following sections simply represents present views that may have to be modified in the light of new and further research. This is one of those areas of life in which 'time makes ancient good uncouth'.

7.12   Obesity

Body mass index (BMI) is used as a measure of weight to height ratio and allows us to recognise four categories of body sizes. The BMI is defined as weight (expressed in kilogram) divided by height squared (expressed in centimetre) and one classification is:

- Underweight   <18.4
- Normal   18.5–24.9
- Overweight   25.0–29.9
- Obese   30.0–39.9
- Severely obese   >40.0

A growing number of persons fall into the last three categories, probably as a consequence of imbalance over many years between increased caloric intake and decreased energy requirement resulting from more sedentary and less active lifestyles. An alternative measure considers the distribution of adipose tissue expressed in a waist to hip ratio. In tackling this problem attention is focussed on fat because it is the most energy dense of our nutrients. It is worth noting that the human race has developed over evolutionary time in a situation where lack of food was a common occurrence and surplus food, over anything but short periods of time, was virtually unknown except for a few rich people. The human system did not need to be designed to deal with the problem of over-consumption.

The problem is not new, even if it is becoming more widespread. Shakespeare has Henry tell Falstaff 'Leave gourmandising. Know that the grave doth gape thrice wider for thee than for other men'.

The problem of obesity is partly genetic and partly environmental (food intake and physical inactivity). Attention is often focussed on long hours spent TV watching where inactivity is often accompanied by poor eating habits. In the USA nearly two-thirds of the population is overweight or obese and almost 40% are clinically obese. Concern is growing about the increase of obesity in children and adolescents.

In 1991 deaths associated with obesity in the USA (300 thousand) were second only to deaths associated with smoking (400 thousand) and it is likely that in the intervening years these numbers have become closer. Obesity is a potent risk factor for type-2 diabetes, hypertension, and dyslipidemia. The average US person seeking treatment for obesity weighs around 100 kg. However, it is worth noting that while fat intake in the USA fell on the basis of percentage of energy from ~42% in the 1960s to ~35% in 1990 fat intake in terms of gram per day fell between the 1960s and the 1980s then stabilised and began to increase in the mid-1990s.

In Europe also, obesity figures are on the increase and it is reported that 20% of men and 25% of women in the UK are obese, that these levels have tripled in the last 20 years, and that 9000 deaths a year are associated with this condition at a cost of £2.5 billion. Obesity is a major public health problem throughout Europe, especially among women in Southern and Eastern European countries. These countries are also among the highest for CVD. In Europe the treatment of obesity-related diseases accounts for 8% of all medical costs.

There are factors other than dietary fat that are important in obesity and attention has been drawn to the beneficial role of dietary calcium in the partitioning of dietary energy, resulting in reduction in body fat and acceleration of weight loss and fat loss during periods of energy restriction. Dairy sources of calcium exert substantially greater effects than supplemental or fortified sources of calcium. This is considered to have important implications in the prevention of paediatric and adult obesity particularly in the light of the marginal calcium intakes exhibited by the majority of the population.

In discussing diet, obesity, and cardiovascular risk Bonow and Eckel (2003) have written: 'The recipe for effective weight loss is a combination of motivation, physical activity, and caloric restriction; maintenance of weight loss is a balance between caloric intake and physical activity with lifelong adherence. For society as a whole prevention of weight gain is the first step in curbing the increasing epidemic of overweight and obesity … physicians should recommend a healthy lifestyle that includes regular physical activity and a balanced diet'. For dietary fat they recommend: total fat 33 energy%, saturated acids 10%, polyunsaturated fatty acids 6% (and not exceeding 10%), *cis*- monounsaturated acids 12%, and *trans* unsaturated acids <2%.

7.13   Coronary heart disease

It has long been known that the blood system is important as a means of moving oxygen and nutrients from cell to cell and from organ to organ and as part of the system by which animals dispose of waste products. This system depends on a complex series of channels of wide-ranging diameter through which blood flows and through which gases and liquids pass from cell to blood stream and vice versa, and of a pump to force the necessary movement. All these have to last for life and any fault with the pump (heart) or the channels (veins and capillaries) will cause difficulty.

CVD is a broad-term embracing diseases of the blood vessels of the heart (CHD), brain (cerebrovascular disease, stroke), and the limbs (peripheral vascular disease). CVD is usually a culmination of atherosclerosis (accumulation of material – plaque – in the walls of arteries of cells comprising connective tissue, lipids, calcium, and debris resulting from cellular breakdown) and thrombosis.

CHD is a major cause of death in the developed world with a peak age of death of 70–74 for men and 75–79 for women but its too-common occurrence at an earlier stage in life is of greater concern. There are three stages in the development of CHD. Initial arterial injury leads to deposition of lipid and cell material (atherosclerosis) and to small blood clots (thrombi) which contribute to the build up of fibrous plaque. Finally, instability of the plaque triggers formation of a major blood clot (thrombus) in the already-narrowed artery. This gives the potential for the blood (and oxygen) supply to the heart muscle to be blocked completely leading to myocardial infarction (heart attack). More simply: the three stages are injury of coronary arteries, fibrous plaque formation, and thrombosis leading to heart attack or stroke. The following have been recognised as risk factors: high blood pressure, high levels of plasma LDL cholesterol, low levels of plasma HDL cholesterol, high levels of plasma fibrinogen, and low levels of plasma antioxidants. These risk factors are linked to a range of controllable and uncontrollable factors. The uncontrollable factors are family history, being male, advancing age, racial origin (Asians show higher rates of incidence than white Caucasians), and possibly low birth weight. Controllable factors include smoking, lack of exercise, stress, and diet. Serum cholesterol level should be below 230 mg/100 ml but very low cholesterol levels (below ~160/180 mg/ml) are also undesirable.

The lipid hypothesis in respect of CHD is concerned with the relationship of blood cholesterol and SFA with CHD mortality. Diets with a high content of fat/SFA/cholesterol lead to high concentrations of total cholesterol in the blood and especially of LDL-cholesterol which results in a high morbidity and mortality from CHD. However, reducing the amount of fat/SAF/cholesterol in the diet reduces the concentration of cholesterol in the blood and especially in the LDL and this results in a lower risk of CHD and eventually a fall in morbidity and mortality. There remains, however, the question as to how far reduced CHD mortality is linked to dietary changes and how far it is related to improved methods of medical treatment.

This hypothesis is the basis of much dietary advice relating to fat consumption though it should be noted that there are those who dispute this proposal and have mounted strong arguments against it (Gurr, 1999). Once the link between cholesterol blood levels and fatty acid intake was accepted it became apparent that blood cholesterol levels could be raised or lowered by dietary changes in fatty acid composition and attempts were made to express these changes in the form of mathematical equations. These predictive equations are related to changes in cholesterol levels and not to absolute values. The earliest of these equations (Keys, 1957) covered only saturated and polyunsaturated acids (expressed as percentage of energy) but later versions include cis-monounsaturated acids, *trans* monounsaturated acids, and some even distinguished between individual fatty acids.

The Keys equation of 1957 showed that a rise in saturated acids led to a rise in cholesterol levels, that polyunsaturated acids had the opposite effect, and that monounsaturated acids were considered to be neutral. This conclusion led to the development of spreads with reduced content of saturated acids and increased levels of linoleic acid, often based on sunflower oil. The coefficients in the equations indicate that the beneficial effect of polyunsaturated acid is only one half the undesirable effect produced by raising the level of saturated acids. More recent equations distinguish three saturated acids and ignore those of shorter- or longer-chain length. Lauric acid has only a minor effect and myristic is greater than palmitic acid. *trans*-Acids also raise cholesterol levels with a slightly greater effect for those from partially hydrogenated fish oils than for those from partially hydrogenated vegetable oils. In a survey of five recent equations the regression coefficients for total cholesterol levels are saturated acids (+0.56 to +0.50), *trans* acids (+0.31 to +0.39), and for polyunsaturated acids (−0.15 to −0.31). The cholesterol-raising effects of the saturated acids are greater than the cholesterol-lowering effects of the polyunsaturated fatty acids.

These equations relate to fatty acids present in the diet as glycerol esters and this requires another factor to be taken into account since there is good evidence that SFA are more atherogenic in the 2-position. For example, tallow and lard both contain about 24% of palmitic acid but differ in that in lard almost all the palmitic acid is in the 2-position. Lard is much more atherogenic than tallow but after interesterification (randomisation) both fats with about 8% of palmitic in the 2-position are equally atherogenic. This effect may be related to the fact that palmitic acid in the 2-position is absorbed more efficiently. Similar effects have been observed with synthetic triacylglycerols and with appropriate vegetable oils.

One study of CHD concludes that fats in the diet should not exceed 33 energy% with saturated acids (10%), polyunsaturated fatty acids (6% and not exceeding 10%), monounsaturated fatty acids (12%), and *trans* acids (<2%) at the levels indicated. This is accompanied by advice to eat more fruit and vegetables, more starch foods, more oily fish, and less full-fat dairy products, fatty meat and meat products, spreadable fats, and high-fat bakery products, to choose low-fat options where possible, and to use less salt. Dietary advice to the general population should always be part of a package that includes advice to stop smoking, to take more exercise, to maintain a healthy weight, and to relieve stress.

The evidence for a beneficial role for long-chain omega-3 polyunsaturated fatty acids is becoming stronger, especially for secondary prevention. Intakes of 800–1000 mg/day are considered to be prudent for those at risk of a secondary attack. At the same time, a high intake of linoleic acid should be discouraged because of its antagonistic effect on the incorporation of omega-3 acids into membranes. The most that can be claimed is that 'consumption of long-chain omega-3 acids may reduce the risk of CHD'. EPA-derived leukotrienes have less potent leukocyte activating effects than AA-derived leukotrienes and at least part of the anti-atherogenic mechanism of the omega-3 acids is likely to be due to their effect on eicosanoid metabolism.

7.14 Diabetes

Diabetes mellitus is a chronic disease in which the metabolism of sugars (and of fats and proteins) is disturbed by a lack of the hormone insulin produced by the endocrine part of the pancreas or by its decreased activity. The main symptom is an increase in the level of blood sugar provoking acute symptoms such as thirst, frequent voiding, and weight loss. The incidence of this disease is increasing all over the world and it is predicted that it will affect 210 million people by 2010. Diabetes is an independent risk factor for CVD.

Type-1 diabetes, representing only about 15% of cases, is found particularly in children, adolescents, and young adults. It results from auto-immune destruction of the insulin-secreting cells of the pancreas. Production of insulin declines and eventually ceases. However, most diabetic individuals (85%) have type-2 diabetes. Two dysfunctions are involved: decreased insulin secretion after a glucose challenge and a decrease in its activity on target organs (liver and muscles). This is called insulin resistance. Obesity is a major pre-disposing factor of this type of diabetes which is largely determined by genetic factors. The metabolic consequences of this defect may not be apparent until the appearance of chronic complications.

One discussion on the nutritional management of this disease suggests that individuals with normal body weight and normal lipid levels should limit fat intake to less than 30% total energy with

SFA restricted to 10%, polyunsaturated acids to less than 10%, and monounsaturated acids at 10–15%. Those with elevated LDL levels should reduce saturated acids to 7% and cholesterol intake to less than 200 mg/day. However, omega-3 polyunsaturated fatty acids should not be restricted. Other aspects of nutritional management of this disease are not included here.

It is known that the various desaturases involved in the conversion of C_{18} polyunsaturated fatty acids to the important acids of longer-chain length such as arachidonic acid, EPA, and DHA are decreased in diabetic patients. As a consequence the phospholipids in tissue lipids contain more saturated and monounsaturated acids and less LCPUFA, especially AA. This, in turn, affects membrane fluidity and eicosanoid production.

7.15 Inflammatory diseases

Inflammation is characterised by swelling, redness, pain, and heat in localised areas of the body. These symptoms result from a series of interactions between cells of the target tissue, cells of the immune system, their products such as eicosanoids, cytokines, and immunoglobulins, and blood components.

Polyunsaturated fatty acids of the omega-3 series act directly by inhibiting AA metabolism and indirectly by altering the expression of inflammatory genes. They are considered to be of therapeutic value for a variety of acute and chronic inflammatory conditions. But what is the appropriate balance between omega-6 and omega-3 acids? This may differ at different parts of the life cycle such as early development and ageing and has yet to be determined.

7.16 Psychiatric disorders

In view of the importance of brain phospholipids and their component acids it is not surprising that the relation between dietary lipids and psychiatric disorders such as schizophrenia and depression have been investigated. Schizophrenic patients are known to have lower levels of polyunsaturated fatty acids (especially linoleic and arachidonic) in their brain phospholipids. This may be the consequence of increased phospholipid hydrolysis and/or decreased incorporation. The role of omega-3 acids and especially EPA is under active consideration.

7.17 Cancer

The possibility of a link between cancer and dietary fat has received intensive study but no consensus has emerged. Such studies are complicated by the fact that cancers in different organs may react differently to dietary fats. Much attention has been given to the possible beneficial treatment of breast cancer with CLA (Section 7.8) but there is still no final conclusion.

Major Edible Uses of Oils and Fats

8.1   Introduction

As already indicated in the previous chapter about 153 million tonnes of 17 commodity oils and fats were produced in 2006/07. It has generally been assumed that around 80% is used for human food (122 million tonnes) but the increasing demand for biodiesel has probably reduced this to a figure close to 113 million tonnes. However, this figure is too high as a measure of fat consumption by reason of loss and waste but also too low because of other fat sources not included in statistical tables for commodity oils. This chapter describes the major food uses of oils and fats. Relevant information on the nutritional properties of these materials is given in the previous chapter.

8.2   Spreads – butter and ghee

Milk and its products must be one of the earliest foods known to man. The dairy industry is based almost entirely on cow milk but there is also a limited supply of milk from goats, camels, and sheep. Cow milk is consumed as a drink (with 2–4% of fat), as cream (30–45%), as butter (82%), and as cheese (30–50%), each of which has the fat level indicated.

Butter is a water-in-oil emulsion consisting of fat (80–82%) and an aqueous phase (18–20%) containing salt and milk-solids-not-fat. The upper legal limit for water is 16%. Butter is made from cow milk (3–4% fat) that is converted first to cream (30–45% fat) by centrifuging and then to butter by churning and kneading. During churning there is a phase inversion from an oil-in-water to a water-in-oil emulsion.

Present annual production of butter is close to seven million tonnes (Table 2.1). The biggest consumers are in the Indian sub-continent (India and Pakistan) and in Europe, particularly in France and Germany.

Butter from cow milk fat has been used as a spread and for baking and frying for many centuries but now it has to compete with good quality spreads. The disadvantages associated with butter are its high price compared to other spreads, its poor spreadability from the refrigerator, and its poor health profile resulting from its high fat content, its high levels of saturated acids and of cholesterol, and the presence of *trans* unsaturated acids. Its advantages are its wholly natural profile and its superb flavour. The name butter is jealously guarded and is legally defined in many countries. It is not permissible to take anything away or to add anything to a product called butter.

Attempts have been made to overcome the above disadvantages by modifying the diet of the cow, by fractionation of anhydrous milk fat (AMF) to give material with modified physical properties (see below), or by blending with an unsaturated vegetable oil. This last improves the healthiness of the product and makes it spreadable from the refrigerator. Products are available in many countries that are blends of butter and vegetable oil – frequently soybean oil. These cannot be called butter since they do not meet the legal specifications but are given an appropriate name that the consumer comes to think of as 'spreadable butter'. Spreadable butters developed in New Zealand are made by fractionation of butter followed by recombination of appropriate fractions.

In times of over-supply there is an interest in extending the range of applications of milk fat by fractionation (Section 3.5). However, the triacylglycerol composition of milk fat is so complex (no individual triacylglycerol exceeds 5%) that differences between crystallised fractions are not so marked as with simpler vegetable oils such as palm oil. Nevertheless, useful separations have been achieved giving fractions that are harder and softer than the original milk fat. The lower melting (softer) fractions are employed to make spreadable butter and the harder fractions find pastry applications. AMF itself is used to make cakes. Mixed with the olein fraction it is used in cookies, biscuits, and butter cream. Mixed with the stearin fraction it is used in fermented pastries and puff pastry. The olein fraction on its own is used in ice cream cones, waffles, butter sponges, and in chocolate for ice cream bars.

In Europe, 'butters' with reduced fat levels (and therefore reduced caloric values) are designated as butter only when they contain 80–90% fat, 'three quarter fat butter' has 60–62% fat, 'half fat butter' has 39–41% fat, and 'dairy fat spreads' have other fat levels. In the United States 'light butter' must contain less than half of the normal level of fat and 'reduced butter' less than one quarter of the normal level.

In India, milk fat is consumed partly as butter but also as ghee. This is a concentrate of butter fat with over 99% milk fat and less than 0.2% moisture. It has a shelf life of 6–8 months even at ambient tropical temperatures. Butter or cream is converted into ghee by controlled heating to reduce the water content to below 0.2%. In other procedures the aqueous fraction is allowed to separate and some of it is run off before residual moisture is removed by heating. The vegetable oil-based alternative to ghee is called vanaspati (Section 8.3).

8.3 Spreads – margarine, vanaspati, and flavoured spreads

Margarine has been produced for more than 100 years. During the 1860s, large sections of the European population migrated from country to town and changed from rural to urban occupations. At the same time, there was a rapid increase in population in Europe and a general recession in agriculture leading to a shortage of butter, especially for the growing urban population. As a consequence, the price rose beyond the reach of many poor people. The situation was so bad in France that the government offered a prize for the best proposal for a butter substitute that would be cheaper and would also keep better.

The prize went to the French chemist, Hippolyte Mège Mouriés, who patented his product in France and Britain in 1869. His process required the softer component from fractionated tallow, skimmed milk, and macerated cow's udder. The product was described as mixed glycerol esters of oleic and margaric acids and was therefore called oleo-margarine. Margaric acid was thought to be heptadecanoic acid (17:0), but it was actually a eutectic mixture of palmitic (16:0) and stearic (18:0) acids. Even this early process involved fractionation and enzymes. Both 'margaric' and 'margarine' should be pronounced with a hard *g* as in Margaret. All three words come from the Greek word for 'pearl' – *margarites*.

Margarine was first considered a cheap and inferior substitute for butter. In several countries regulations were passed that prohibited the addition of colouring matter so that white margarine would compare even less favourably with the more familiar yellow butter. Now the situation is different. These impediments have largely disappeared and margarine is widely accepted as having several advantages over butter. It is a more flexible product that can be varied for different markets and modified to meet new nutritional demands, such as desirable levels of cholesterol, phytosterols, saturated or *trans* acids, and fat content, as well as the statutory levels of certain vitamins. Table margarine is made from appropriate oils and fats (soybean, rapeseed/canola, sunflower, cottonseed, palm, palm kernel, coconut), which may have been fractionated, blended, hydrogenated in varying degrees, and/or interesterified. Fish oil (hydrogenated or not) may also be included. Other ingredients include surface-active agents, proteins, salt, and water along with preservatives, flavours, and vitamins.

Margarine production involves three basic steps: emulsification of the oil and aqueous phases, crystallisation of the fat phase, and plasticification of the crystallised emulsion. Water-in-oil emulsions are cooled in scraped-wall heat exchangers during which time fat crystallisation is initiated, a process known as 'nucleation', during which the emulsion drop size is reduced. There follows a maturing stage in working units during which crystallisation approaches equilibrium, though crystallisation may continue even after the product has been packed. The lipid in margarine is part solid (fat) and part liquid (oil), and the proportion of these two varies with temperature. The solid/liquid ratio at different temperatures is of paramount importance in relation to the physical nature of the product.

Individual crystals are between 0.1 and several micrometres in size and form clusters or aggregates of 10–30 μm. One gram of fat phase may contain up to 10^{12} individual crystals. The aqueous phase is present in droplets, generally 2–4 μm in diameter, stabilised by a coating of fat crystals.

It is desirable that margarine taken from the refrigerator at 4°C should spread easily. For this to happen the proportion of solids should be 30–40% at that temperature and should not exceed the higher value. For the sample to 'stand up' at room temperature (and not collapse to an oily liquid) it should still have 10–20% solids at 10°C. Finally, so that it melts completely in the mouth and does not have a waxy mouth feel, the solid content at 35°C should be less than 3%. These are important parameters that can be attained with many different fat blends. Formulations have to be changed slightly to make the product suitable for use in hot climates.

Fats usually crystallise in two different forms, known as β' and β. The β form is thermodynamically more stable and will therefore be formed in many fats and fat blends. But sometimes the fat remains in the slightly less stable β' form. The β' form is preferred for margarines and other spreads because the crystals are smaller, are able to trap more liquid to give firm products with good texture and mouth feel, and impart a high gloss to the product. The β crystals, on the other hand, start small but tend to agglomerate and can trap less liquid. It is therefore desirable to choose a blend of oils that crystallise in the β' form.

Margarines and shortenings made from rapeseed/canola, sunflower, and soybean oil after partial hydrogenation tend to develop β crystals. Stable β' crystals are formed through incorporation of some cottonseed oil, hydrogenated palm oil or palm olein, tallow, modified lard, or hydrogenated fish oil. The canola, sunflower, and soybean oils all have very high levels of C_{18} acids, whereas the remainder have appreciable levels of C_{16} acids (or other chain length in the case of fish oil) along with the C_{18} acids and thus contain more triacylglycerols with acids of mixed chain length.

To make spreads (and shortenings) from readily available liquid vegetable oils it is necessary to 'harden' them (*i.e.* increase the solid/liquid ratio). This requires that the proportion of solid triacylglycerols be increased and for most of the last 100 years this has been achieved by partial hydrogenation that converts linoleic acid to saturated acids and to monoene acids rich in *trans* isomers (Section 3.6). Since the latter are higher melting than their *cis* isomers this was seen as an additional route to solid compounds. However, during the later years of the 20th century researchers in Europe showed that *trans* acids have greater cholesterol-raising powers than do the saturated acids. One country in Europe (Denmark) prohibited the use of fats and fatty ingredients with *trans* content above 2% and food producers in other European countries developed recipes with lower levels of *trans* acids. This could result in the use of more saturated acids but the combined content of saturated and *trans* acids was lowered. Changes in the USA were spurred by legislation, operative from January 2006, requiring separate labelling of *trans* acids (excluding conjugated linoleic acid, CLA). A product can be labelled as *trans*-free only if the content of *trans* acids is less than 0.5 g per serving of the food product and many food companies in the USA changed their recipes to deliver *trans* acids below this limit. They are then able to claim *trans*-free products. This has been achieved, in part, by optimising the partial hydrogenation procedure to minimise (but not eliminate) *trans* acids and also by a new approach in which unhydrogenated oil is blended with hardstock and the mixture subjected to interesterification. The hardstock may be a lauric oil, palm stearin, or a fully hydrogenated oil. Since this last contains very little unsaturated acid *trans* acids must be virtually absent. A problem may remain if it is necessary to declare the presence of hydrogenated vegetable oil because of the unwarranted perception that this is undesirable. In some quarters it is feared that one day chemical interesterification may also be perceived to be unacceptable because of the use of 'chemicals'. If that happens suppliers will have to use enzymatic interesterification. This has some advantages but is more costly than chemical interesterification (Section 3.8).

It is impossible to list all of the formulations used to make spreads, and the following list is merely indicative (in the following blends, 'hydrogenated' means 'partially hydrogenated'). These blends are now being modified along the lines indicated in the previous paragraph to give lower levels of *trans* acids.

- Blends of hydrogenated soyabean oils with unhydrogenated soybean oil.
- Blends of canola oil, hydrogenated canola oil, and either hydrogenated palm oil or palm stearin.
- Blends of various hydrogenated cottonseed oils.
- Blends of edible tallow with vegetable oils (soybean, coconut).
- Blends of palm oil with hydrogenated palm oil and a liquid oil (rapeseed, sunflower, soybean, cottonseed, olive).

For hot climates a harder formulation is required, as in the following examples from Malaysia:

- Palm oil (60%), palm kernel oil (30%), and palm stearin (10%).
- Palm stearin (45%), palm kernel oil (40%), and a liquid oil (15%).

Table 8.1 gives details of the fatty acid composition of butter and of soft tub margarine and Table 8.2 provides information on production levels of margarine in the period 2001–2006.

Spreads are expected to have a shelf life of about 12 weeks. With good ingredients and the absence of pro-oxidants (e.g. copper), oxidative deterioration is not likely to be a problem. However, care must be taken to avoid microbiological contamination in the aqueous phase. This is achieved by hygienic practices during manufacture, the addition of some salt (8–10% in the aqueous phase, corresponding to slightly more than 1% in the margarine), control of pH of any cultured milk that may be used, and careful attention to droplet size in the emulsion.

The levels of total *trans* acids (mainly 18:1 but also some 18:2 and 18:3) in margarines from various countries are listed in Table 8.3. Levels have declined over the last 10 years and are now noticeably

Table 8.1 Approximate fatty acid composition of spreading fats (%)

Fat	Saturated	Monoene*	Polyene
Butter	63–70	28–31	1–3
Margarine (soft, tub)	17–19	35–52	29–48

*Soft margarines traditionally contained 10–18% *trans* acids but products with lower levels of *trans* acids are now being produced.

Table 8.2 Production of margarine (million metric tonnes) expressed in terms of normal fat levels and including vanaspati in the period 2001–2006

Country	2001	2002	2003	2004	2005	2006
World	9.74	9.84	9.66	9.73	9.94	10.07
EU-27	2.66	2.66	2.55	2.50	2.48	2.51
Pakistan	1.43	1.55	1.57	1.58	1.61	1.63
CIS	0.93	0.94	1.03	1.12	1.23	1.27
India	1.40	1.43	1.21	1.22	1.10	1.13
Turkey	0.49	0.51	0.54	0.54	0.62	0.64
USA	0.82	0.77	0.73	0.72	0.64	0.66
Brazil	0.49	0.49	0.49	0.49	0.49	0.50
Other	1.52	1.49	1.54	1.56	1.77	1.73

Note: The reduction in Indian production results from increased imports from Sri Lanka and Nepal following new tariff agreements.

It has been estimated that production of vanaspati in 1998 was 4.7 MMT (mainly in Pakistan 1.4, India 1.0, Iran 0.5, and Egypt 0.4).

Table 8.3 Presence of *trans* fatty acids in margarine

Country (year of publication)		Mean (%)
Germany (1997)	0.2–5	1.5
Belgium, Hungary, and Britain (1996)	1–24	9.7
Denmark (1998), soft		0.4
Denmark (1998), hard		4.1
Canada (1998), tub	1–46*	18.8
Canada (1998), hard	16–44	34
Hawaii (2001), cup	1–19	12.1
Hawaii (2001), carton	18–27	23.4

*Mainly 15–20%.
Source: Adapted from Gunstone in Akoh and Min (2008).

lower than those cited. However, spreads are not the only source of dietary *trans* fatty acids. Such acids are also obtained from dairy produce and from baked goods made with partially hydrogenated vegetable fats. Ratnayake *et al.* reported in 1998 that with a *trans* fatty acid consumption of about 8.4 g/day in Canada, only about 0.96 g (11%) comes from the consumption of margarine. The balance comes from fast foods, bakery goods, and ruminant fats. In 2000, Wolff *et al.* drew attention to the very different profile of *trans* monoene fatty acids consumed in France and Germany compared with consumption in North America. These differences reflect the differing nature of *trans* acids from dairy produce on the one hand and industrially hydrogenated vegetable oils on the other.

Spreads are now available with added phytosterols which are claimed to reduce blood cholesterol levels. The phytosterols, added at around the 8% level, are obtained either from tall oil and added to spreads as hydrogenated sterol esters (stanols) or from soybean oil and added as unsaturated sterol esters. Spreads are suitable foods for phytosterol addition because they are used widely and regularly but are unlikely to be over-consumed. Intake of phytosterols is normally 200–400 mg/day, though higher for vegetarians, but the intake of 1.6–3.3 g/day, recommended by those offering this special margarine, is markedly higher. Normally about 50% of ingested cholesterol is absorbed but with an adequate intake of phytosterols, which are absorbed only at the 5% level, absorption of cholesterol falls to about 20%.

Spreads with reduced levels of fat (40% or less) are popular with consumers (as an alternative to discipline in the amount of normal spread consumed). These spreads contain more water than the full-fat spreads and require emulsifiers (monoacylglycerols or poly-glycerol esters). It is also usual to add thickeners, such as gelatin, sodium alginate, pectin, and carrageenan, to the aqueous phase. Industrial margarines are used mainly for bakery products and are discussed in Section 8.4.

Chocolate spreads are increasingly popular. They consist of a soft smooth fat (generally at about a 30% level) and cocoa powder and may also contain nuts. Like other spreads there must be an appropriate mixture of solid and liquid fats. They are designed to be kept at room temperature (but are often kept in a refrigerator) and to be spread on bread, toast, or biscuits. The spread smells and tastes like chocolate but does not solidify. To prevent oiling out (separation of oil) they should be made with fats crystallising in the β' form. Traditionally partially hydrogenated fats have been used but attempts are now being made to develop formulations containing less *trans* acids.

Vanaspati can be considered as vegetable ghee and is used mainly for frying and for the preparation of sauces, sweets, and desserts. Traditionally, vanaspati was a blend of hydrogenated seed oils (cottonseed, groundnut, soybean, rapeseed/canola, and palm), but increasingly palm oil has become a significant component. The product should melt between 31°C and 41°C, though generally it is close to 37°C in India and is 36 $\pm$2°C in Pakistan. Because of the method of production involving hydrogenation, vanaspati contains high levels of acids with *trans* unsaturation (more than 50% in India and about 27% in Pakistan) but with increasing use of palm oil in vanaspati the need for hydrogenation is reduced with a consequent fall in the level of *trans* acids. Figures around 3% have been reported in Pakistan.

8.4 Baking fats and shortenings

The use of oils and fats in baking processes ranks with frying and spreads as a major food use of these materials. The products range from breads and layered doughs to cakes, biscuits (cookies) and biscuit fillings, pie-crusts, short pastry, and puff pastry. The fats used to produce this wide range of baked goods vary in their properties and particularly in their melting behaviour and plasticity.

It is possible to attain appropriate properties with different blends of oils, and preferred mixtures vary in different regions of the world. In addition to the desired physical properties, it is necessary to meet two further requirements. One is oxidative stability related to the shelf life of the baked goods. The other is the need to respond to current nutritional demands. A good baked item will be tasty, have good texture, have a reasonable shelf life in terms of rancidity and palatability and texture, and will be a healthy food. Sometimes the pressure for appropriate physical properties and nutritional requirements work in opposite directions and a compromise has to be made. As already discussed with the spreads, a plastic fat containing solid and liquid components must have some solid triacylglycerols, which implies a certain level of saturated acids or of acids with *trans* unsaturation despite the nutritional concerns associated with these compounds.

Fats used to make doughs of various kinds are almost entirely plastic fats, that is mixtures of solid and liquid components that appear solid at certain temperatures but deform when a pressure is applied. Fats exert their influence by interaction with the flour and (sometimes) sugar, which are the other major components of a baked product.

Baking fats may include butter or margarine both of which are more than 80% fat and also contain an aqueous phase, or they may be shortenings with 100% fat. These are so described as they give pastry the crispness and flakiness that is suitable for its edible purpose. Industrial margarine has the fat/water ratio required of margarine but differs from the domestic spread in that it has fat components selected specifically to produce the physical properties required by its final end use. Changes in the composition of fat in margarines and spreads designed to increase their nutritional value have not always carried through to the baking fats, which are often richer in saturated fatty acids and/or acids with *trans* unsaturation. But there seems little doubt that the appropriate changes will come. Baked goods contain what is described as 'hidden' fat, and it is easy to forget the presence of fat when delicious pastries, cakes, and biscuits are being eaten.

The prime function of fat in a cake is to assist in aeration and to modify the texture of the product. The first stage in making a cake is to produce a batter containing a fine dispersion of air bubbles largely stabilised by fat crystals. During baking the fat melts and the water-in-oil emulsion inverts with the air being trapped in the aqueous phase. As baking continues the starch is hydrated and gelatinised, the protein starts to coagulate, and the air cells expand through the presence of steam and carbon dioxide (produced from baking powder).

In short pastry, aeration is only of secondary importance. The fat needs to be fairly firm and should be distributed throughout the dough as a thin film: lard, beef tallow olein, and hardened vegetable oils may be employed. Sometimes butter or margarine is used.

In puff pastry (pie-crusts, Danish pastries, croissants), fat acts as a barrier separating the layers of dough from one another. Liberation of gas or steam during baking produces a layer structure. This requires a fat of higher melting point fat than normal (about 42°C) with a higher solid fat content achieved through an appropriate degree of hydrogenation. Small amounts of fat (2–5%) are added to bread dough.

8.5   Frying oils and fats

The use of oils and fats as a frying medium in both shallow and deep frying mode is an important component in the whole picture of food applications. Pre-fried and fried foods are now a significant component of our dietary intake and it is reported that more than 20 million tonnes of oils and fats are used in this way. Since some of this is discarded after use, not all of it is consumed.

Frying is usually carried out at a temperature of 165–185°C and is an efficient method of heat transfer that allows quick cooking and adds flavour to fried food. Some oil is absorbed by the fried food. In shallow pan frying surplus oil is cleaned away at the end of the frying operation. In deep fat frying oil is re-used until it has to be discarded because of its poor quality. This low-grade material may be added to animal feed as an energy source or it may be converted to methyl esters and used as biodiesel.

During frying there is heat and mass transfer. When frying oil is maintained at 150–180°C within the food there is a dried food zone (103–150°C) closest to the hot oil, a vaporisation region (100–103°C), a migration region (100°C), and at the centre a liquid water region (75–100°C). There is movement of water (as steam) out of the food into the hot oil and some movement of fat into the food. With well-cooked food these changes produce a crisp outer layer in which there have been changes to carbohydrate and protein. It is desirable to minimise oil penetration occurring both during cooking and during cooling. If the frying temperature is too low cooking is slow and more fat is incorporated into the food, if it is too high then undesirable changes in the oil are accelerated.

Frying is carried out on a domestic scale, in restaurants and fast food outlets on a batch scale using 4–20 kg of oil, and under industrial conditions in continuous mode with one tonne or more of oil to produce fried products for retail outlets. Popular fried foods include French fries, chicken, fish, meat, potato crisps, tortilla chips, extrusion snacks, doughnuts, nuts, and noodles. During frying, oil is transferred to the food so that fried foods contain additional fat at a level of 10–40%. Fat from the food is also transferred to the frying oil so that though oil quality is controlled at the beginning of the frying process the oil soon becomes contaminated, for example, with fish oil or with animal fat depending on the food being fried.

The frying process, conducted at elevated temperature and in the presence of air, results in several chemical changes in the frying oil. Oxidation is accompanied by dimerisation, polymerisation, and fatty acid cyclisation. It also produces short-chain carbonyl compounds such as decadienal from linoleate which give a distinctive flavour to the fried product (Section 6.2). Volatile products are quickly lost through steam distillation which occurs during frying and accounts for the characteristic smell of frying operations but compounds of higher molecular weight remain in the frying oil. With continued use the oil begins to smoke, to foam, and to become more viscous. Oil absorbed by the fried food has to be replaced by fresh oil and turnover and replacement of fat are important factors in a good quality frying operation. This ensures that low-quality oil is not being used. Under the best frying conditions the major health concern may not be the small amount of artefacts but rather the increased level of fat that is being consumed.

A good frying oil will have high oxidative stability, a high smoke point (corresponding to a low level of free fatty acid), and show minimum colour darkening during use. The oil may be chosen because it gives a distinctive flavour to the fried food as with corn oil, olive oil, groundnut oil, and tallow. Alternatively, a refined blend may be used (cottonseed, groundnut, soybean, palm olein) which may have been subject to partial hydrogenation.

On nutritional grounds the ideal frying oil contains low levels of saturated acid and of unsaturated acids with *trans* unsaturation.

Low levels of polyunsaturated fatty acids are also preferred since these are the precursors of undesirable oxidised and polymerised products. Consequently there should be a high level of oleic glycerol esters. In practice it may not be possible to attain all these objectives. Differing frying media are used in different parts of the world depending on local availability and on cultural preferences for particular flavours. They include lauric oils, palm oil and palm olein, groundnut oil, rice-bran oil, cottonseed oil, corn oil, sunflower oil, soybean and rapeseed oils after brush hydrogenation to reduce the level of linolenic acid, and a range of high-oleic varieties. In France and Belgium it is forbidden by law to use frying oils with more than 2% of linolenic acid. Special oils used for frying include olestra (polyacylated sucrose with virtually zero calorific value because it is not absorbed) in the USA and 'Good Fry' in Europe. The latter is a high-oleic sunflower oil with up to 6% of sesame oil and/or rice-bran oil, both of which contain powerful natural antioxidants (Section 6.2). Many high-oleic oils are being developed through conventional seed breeding or through genetic modification and all these will be considered as potential frying oils (Table 8.4).

Frying oils are monitored by measuring properties such as: acid value, peroxide value, anisidine value, conjugated diene, total polar materials (TPM), and polymeric triglycerides (PTG) (see Chapter 4). The oil should be replaced when these parameters are above specification. It has been recommended that TPM should not exceed 24% and PTG should not exceed 12%. In most European countries frying oil should be discarded when TPM and PTG together exceed 24–27%. It would be better if these standards could be made more uniform. These measurements require laboratory procedures and a number of simple instrumental methods suitable for use in a food-processing factory have also been examined. It must be remembered that oxidative deterioration starting during food processing can continue during storage even at $-5°C$ to $+10°C$ (Stier, 2004).

The discovery of low levels of acrylamide in French fries and crisps and in baked and roasted foods raised some concern and a good deal of investigation since this compound is carcinogenic and has other undesirable physiological properties at higher concentrations. The acrylamide molecule ($CH_2=CHCONH_2$) contains three carbon atoms and it is still not certain whether these come from glycerol (possibly via monoacylglycerols), from amino acids, from reducing sugars or from all these sources. Levels in French fries after frying are around 600 µg/kg (*i.e.* parts per billion) and

Table 8.4 Fatty acid composition of non-hydrogenated oils used in frying

Oil	16:0	18:0	18:1	18:2	18:3
Soybean					
Normal	10.4	4.1	22.9	52.9	7.5
High-oleic	6.4	3.3	85.6	1.6	2.2
Low-saturated	4.3	2.9	19.7	61.8	8.6
Low-linolenic	10	5	41	41	2
Rapeseed					
Normal	4.0	2.0	58.0	20.0	9.0
High-oleic	3.6	2.3	78.8	5.1	5.2
Low-linolenic	4.0	2.0	63.0	23.0	4.0
Sunflower					
Normal	7.0	4.5	18.7	67.5	Trace
High-oleic	3.6	4.3	82.2	9.9	Trace
Mid-oleic	4.6	4.2	61.3	27.2	Trace
Corn oil	10.9	2.0	25.4	59.6	1.0
Cottonseed oil	21.6	2.6	18.6	54.4	0.7
Palm oil	42.9	4.6	39.3	10.7	0.4
Palm olein	39.8	4.4	42.5	11.2	0.4

Source: Adapted from Goetz (2006) Developments in frying oils, in *Modifying Lipids for Use in Foods* (editor F.D. Gunstone), Woodhead Publishing, Cambridge, England, p. 525.
Note: Brush hydrogenation of soybean oil and rapeseed oil will give products with about half these levels of linolenic acid with more 18:1 having *trans* unsaturation. Alternative data are to be found in Table 2.5.

900–1000 µg/kg in crisps. These levels increase with temperature (another reason for frying at the lowest practicable temperature) and change with the variety of potato used.

8.6   Salad oils, mayonnaise and salad cream, French dressing

Salad oils, used to make mayonnaise and salad cream, should be oxidatively stable and free of solids even when stored in a refrigerator at about 4°C. Several vegetable oils may be used. Those containing linolenic acid (soybean oil, rapeseed/canola oil) are usually lightly hydrogenated (brush hydrogenation) to enhance oxidative stability. All oils are generally winterised (Section 3.5) to remove high-melting glycerol esters that would crystallise, as well as waxes present in some solvent-extracted oils. The latter lead to a haze in the oil when it is cooled. Salad oils must pass a 'cold test', which requires that the oil remain clear for 5.5 h at refrigeration temperature. After appropriate treatment, soybean, rapeseed/canola, corn, and sunflower oils are used to produce mayonnaise.

Mayonnaise is an oil-in-water emulsion containing between 65% (legal minimum) and 80% of oil. The aqueous phase contains vinegar, citric acid, and egg yolk. This last contains lecithin, which serves as an emulsifying agent. Lemon and/or lime juice, salt, syrups, seasonings, spices, and antioxidants are optional constituents. These components may be mixed together at temperatures not exceeding 5°C (cold process) or at temperatures around 70°C (hot process). A typical mayonnaise contains vegetable oil (75–80% by weight), vinegar (9.4–10.8%), egg yolk (7.0–9.0%), and small amounts of sugar, salt, mustard, and pepper. 'Light' mayonnaise contains only 30–40% of oil and in low-calorie dressings the level is 3–10%.

Salad creams are similar but contain much less oil (30–40%) along with cooked starch materials, emulsifiers, and gums to provide stability and thickness. They are cheaper than mayonnaise.

French dressings are temporary emulsions of oil, vinegar or lemon juice, and seasonings. Because the emulsions are not stable the dressings should be shaken before use. A non-separating product can be made by addition of egg yolk or other emulsifying agent.

8.7 Chocolate and confectionery fats

Chocolate is an important fat-containing food based mainly, but not always entirely, on cocoa butter. Confectionery fats have similar physical/functional properties but in the EU they can only be described as chocolate if fats other than cocoa butter come from a prescribed list (Table 2.3, Section 2.3) and do not exceed 5% of the final product.

Cocoa beans contain 50–55% fat. Harvested pods are broken open and left in heaps on the ground for about a week during which time the sugars ferment. The beans are then sun-dried and are ready for transportation and storage. To recover the important components the beans are roasted (~150°C), shells are separated from the cocoa nib, and the latter is ground to produce cocoa mass. When this is pressed it yields cocoa butter and cocoa powder still containing some fat. Typically, 100 g of beans produce 40 g of cocoa butter by pressing, expelling, or solvent extraction, 40 g of cocoa powder remaining after extraction with 10–24% fat, and 20 g of waste material (shell, moisture, dirt, etc.). Increasingly the beans are processed in the country where they grow and cocoa liquor, cocoa powder, and cocoa butter (usually in 25 kg parcels) are exported to the chocolate-producing countries.

Both cocoa butter (a solid fat melting at 32–35°C) and cocoa powder are important ingredients in chocolate. Cocoa butter is in high demand because its characteristic melting behaviour gives it properties that are significant in chocolate. At ambient temperatures it is hard and brittle, giving chocolate its characteristic snap. Also, it has a steep melting curve that permits complete melting at mouth temperature. This gives a cooling sensation and a smooth creamy texture. Typically the content of solid falls from 45% to 1% between 30°C and 35°C. The hardness of cocoa butter is related to its solid fat content at 20°C and 25°C. The melting behaviour is linked to the chemical composition of cocoa butter. The fat is rich in palmitic (24–30%), stearic (30–36%), and oleic acids (32–39%), and its major triacylglycerols are of the kind SOS where S represents saturated acyl chains in the 1- and 3-positions and O represents an oleyl chain in the 2-position. There are three major components: POP, POSt, and StOSt (P = palmitic acid and St = stearic acid). Cocoa butter has a high content of saturated acids, which raises health concerns. However, it has been argued that much of this is stearic acid that is not considered to be cholesterolemic. Cocoa butter is also a rich source of flavonoids which have powerful antioxidant activity.

Cocoa is grown mainly in West Africa, South East Asia, and South and Central America. The composition of cocoa butter from these different sources varies slightly. Typical figures for cocoa butter from Ghana are shown in Table 8.5. Small differences in fatty acid composition are reflected in the iodine value and the melting point but more significantly in the triacylglycerol composition and the melting profile. For example, the content of the important SOS triacylglycerols varies between 87.5% in Malaysian and 71.9% in Brazilian cocoa butter, with the African samples midway between these extremes. There is some evidence that the cocoa butters of different geographic origin are becoming more alike.

The crystal structure of cocoa butter has been studied intensively because of its importance in understanding the nature of chocolate. The solid fat has six crystalline forms designated I–VI

Table 8.5 Composition and properties of Ghanain cocoa butter

Iodine value	35.8
Melting point (°C)	32.2
Diacylglycerols	1.9
Free acid (%)	1.5
Component acids	
Palmitic	24.8
Stearic	37.1
Oleic	33.1
Linoleic	2.6
Arachidic	1.1
Component triacylglycerols	
Trisaturated	0.7
Monounsaturated	84.0
POP	15.3
POSt	40.1
StOSt	27.5
Diunsaturated	14.0
Polyunsaturated	1.3
Solid content (pulsed NMR) tempering for 40 h/26°C	
20°C (%)	76.0
25°C (%)	69.6
30°C (%)	45.0
35°C (%)	1.1

Source: Adapted from Shukla (1997) Chocolate – The chemistry of pleasure. *INFORM*, **8**, 152.
Note: The original paper contains more details along with information on cocoa butter from India, Nigeria, and Sri Lanka.

(Section 6.3). Transition from form V to the more stable form VI leads to the appearance of white crystals of fat on the surface of the chocolate. This phenomenon is termed 'bloom'. It is promoted by fluctuations in temperature during storage and by migration of liquid oils from nut centres. It is a harmless change but is considered undesirable because it may be mistaken for microbiological contamination. Bloom can be inhibited by addition of a little 2-oleo 1,3-dibehenin (BOB) to the cocoa butter (B = behenic acid, 22:0). A recent review of this topic is that by Lonchampt and Hartel (2004).

The simplest plain chocolate contains sugar and cocoa liquor, with cocoa butter the only fat present. A typical plain chocolate has cocoa mass (~40%, which contains some cocoa butter), sugar (~48%), added cocoa butter (~12%), and small amounts of lecithin and other materials. Other chocolates may contain higher levels of cocoa butter, generally in the range 25–35%. In EU countries it is permissible to replace cocoa butter with up to 5% of another fat with similar fatty acid and triacylglycerol composition taken from a prescribed list of tropical fats (Section 2.3). The permitted tropical fats come from palm, illipe, shea, sal, kokum, and mango and may be used in a fractionated form (Table 2.3). Hydrogenation and interesterification procedures are not permitted in the preparation of these fats.

Milk chocolate contains less cocoa butter and between 3.5% and 9% of milk fat. White chocolate is based on sugar and cocoa liquor and cocoa butter (without cocoa mass). If the latter is not entirely refined it will retain some of the flavour normally associated with chocolate. Chocolate normally contains up to 0.4% of lecithin to aid the processing of the chocolate by reducing the viscosity of molten chocolate. Polyglycerol ricinoleate is sometimes added to optimise viscosity. Ricinoleic acid (12-hydroxyoleic) is the major acid in castor oil.

Cocoa butter alternatives (CBA) is a general name covering cocoa butter equivalents (CBE), cocoa butter improvers (CBI), cocoa butter replacers (CBR), and cocoa butter substitutes (CBS).

CBE have the same general chemical composition and hence the same physical properties as cocoa butter and include the tropical oils described above and sometimes designated as hard butters. These can be blended to give mixtures of POP, POSt, and StOSt very similar in composition to cocoa butter and fully miscible with it. The level at which cocoa butter can be replaced by a CBE is limited in some countries on a legal basis and not on a functional basis. CBE must be compatible with cocoa butter by virtue of their similar fatty acid and triacylglycerol composition. They have a melting range equivalent to that of cocoa butter, yield the β polymorph when processed and tempered in the same way as cocoa butter, and give a product that is at least as good as cocoa butter with respect to bloom.

CBR are usually based on vegetable oils such as soybean, cottonseed, or palm that have been fractionated or partially hydrogenated. They contain *trans* unsaturated acids at levels up to 60% and have a different triacylglycerol composition from cocoa butter. They do not require tempering but should be compatible with cocoa butter.

CBS are usually based on lauric fats. They share some of the physical properties of cocoa butter but have a different composition. Coatings based on CBS fats do not require to be tempered and are used in the molten state for enrobing. They give a superior gloss and have very sharp melting characteristics.

Chocolate spreads are increasing in popularity (Section 8.3).

8.8 Ice cream

The annual production of ice cream in the United States in 2002 was reported to be about 54 million hectolitres (*i.e.* 5400 million litres), suggesting that the global figure was then at least twice that level. This quantity of ice cream will contain around 0.8–1.0 million tonnes of fat. Traditionally this has been mainly milk fat along with some vegetable fat which might be sunflower, groundnut, palm, palm kernel, and coconut. Some of the newer non-dairy recipes with ~8% fat have lower levels of saturated (around 40%) and *trans* acids (zero).

Ice cream contains water (60–70%) and total solids (30–40%), with the latter including fat (5–12%), milk solids other than fat (10–12%), sucrose (12–14%), glucose solids (2–4%), emulsifier (0.2–0.5%), and stabiliser (0.1–0.3%). Legal requirements for fat vary from country to country as does the possibility of replacing some or all the dairy fat with vegetable fat.

Fat in ice cream contributes to structure. It stabilises the aerated foam, improves melting resistance, imparts creaminess, and contributes to taste. Its most important properties are melting characteristics, solid-to-liquid ratio at various temperatures, and its taste profile.

Production of ice cream occurs through nine stages: selection and weighing of ingredients, mixing of these in an appropriate sequence at 20–35°C, pasteurisation at 70–75°C or sterilisation at 95°C, homogenisation at 75°C, cooling to less than 5°C, ageing at 5°C for at least 4 h, freezing at −5°C to −10°C, hardening at −25°C to −35°C, and storage at −18°C to −20°C.

8.9 Incorporation of vegetable fats into dairy products

Vegetable oils may be incorporated into dairy products as a replacement for dairy fat. This happens when local supplies of milk fat are inadequate as in some tropical countries where the climate is not suitable for large-scale dairy farming and also for consumers concerned about the saturated acids and cholesterol present in milk fat. In addition, it is possible to produce milk fat replacements in a more convenient form as, for example, in long-life cream. The use of vegetable fat in ice cream has already been discussed in Section 8.7.

So-called filled milk is made from skim milk powder reconstituted with an appropriate vegetable oil. This latter should be free of linolenic acid, have a low content of linoleic acid, and contain antioxidant so that it is oxidatively stable. Palm oil, palm kernel oil, and coconut oil are most frequently used, and these may be partially hydrogenated to provide further stability against oxidation.

Non-dairy coffee whiteners, available in powder or liquid form, generally contain 35–45% fat, which is usually partially hydrogenated palm kernel oil.

Cheeses have been developed based on vegetable fat rather than dairy fat. Several formulations have been described incorporating soybean oil with or without hydrogenation, palm oil, rapeseed oil, lauric oils, and high-oleic sunflower oil. Attempts have been made to incorporate into these products, some of the short-chain acids that are characteristic of milk fat and give cheese some of its characteristic flavour.

Non-dairy whipping creams, made with hardened palm kernel oil and coconut oil (each about 17%), are convenient because they have a long shelf life at ambient temperature. First produced for the bakery and catering market with high overrun (increased volume when whipped) and good shape retention, they are now supplied to the retail market for domestic use. Pouring creams, containing about 9% of each of the two lauric oils, are also available. Both creams also contain buttermilk powder (7%), guar gum (0.10–0.15%), emulsifying agent (0.30–0.35%), β-carotene (0.25%), and water.

8.10   Edible coatings

Foods are sometimes coated with thin layers of edible material to extend shelf life by minimising moisture loss, to provide gloss for aesthetic reasons, and to reduce the complexity and cost of packaging. The thin layers may be carbohydrate, protein, lipid, or some combination of these. The lipids most commonly used are waxes (candelilla, carnauba, or rice bran), appropriate triacylglycerols, or acetylated monoacylglycerols. The latter are capable of producing flexible films at temperatures below those appropriate for the waxes even though they are poorer moisture barriers. The foods most frequently coated are citrus fruits (oranges and lemons), deciduous fruits (apples), vegetables (cucumbers, tomatoes, potatoes), candies and confectioneries, nuts, raisins, cheeses, and starch-based products (cereals, doughnuts, and ice cream cones and wafers).

Vegetable oils used to coat food products must be liquid at room temperature and must have high oxidative stability. They serve as a moisture barrier, a flavour carrier, a lubricant or release agent, as an anti-dust or anti-caking agent, and as a gloss enhancer. They are used at low levels and are sprayed on to large exposed surfaces of products during roasting, frying, or handling. Traditionally they are made from commodity oils like soybean or cottonseed. These oils are cheap but require elaborate processing (partial hydrogenation and fractionation) to develop the required physical state and chemical stability. New high-oleic oils may also be used. These are more costly but they bring added value in terms of their superior nutritional properties resulting from lower levels of *trans* acids and saturated acids and in the reduced need for processing. Lauric oils, such as coconut oil, palm kernel oil, are used to spray cracker-type biscuits to provide an attractive appearance, maintain crispness by acting as a barrier to moisture, and improve eating quality.

8.11   Emulsifying agents

Fatty acids and their derivatives are amphiphilic. This means that their molecules have hydrophilic and lipophilic regions. If these are appropriately balanced, then the molecules can exist in a physically stable form between aqueous and fatty substances. They can therefore be used to stabilise both oil-in-water and water-in-oil emulsions and are important components of many of the fat-based products that have been described in the earlier sections of this chapter. Applications of emulsifiers in foods include film coatings, stabilising and destabilising emulsions, modification of fat crystallisation, dough strengthening, crumb softening, and texturisation of starch-based foods. Production of food emulsifiers was estimated to be 250,000 metric tonnes in 1994 of which about 75% is monoacylglycerols or compounds derived from these.

Monoacylglycerols are most often made by glycerolysis of natural triacylglycerol mixtures in the presence of an alkaline catalyst at 180–230°C for 1 h. Fat and glycerol (30% by weight) will give a mixture of monoacylglycerols (around 58%, mainly the 1-isomer), diacylglycerols (about 36%), and triacylglycerols (about 6%). This mixture can be used in this form or it can be subjected to high-vacuum thin-film molecular distillation to give a monoacylglycerol product (around 95% and at least 90% of the 1-monoester) with only low levels of diacylglycerols, triacylglycerols, and free acids. Attempts are being made to develop an enzyme-catalysed glycerolysis reaction that occurs under milder reaction conditions. The oils most commonly used include lard, tallow, soybean, cottonseed, sunflower, palm, and palm kernel oil – all in hydrogenated or non-hydrogenated form. Glycerol monostearate (GMS) is a commonly used product of this type.

The properties desired in a monoacylglycerol for some specific use may be improved by acylation of one of the free hydroxyl groups by reaction with acid (lactic, citric) or acid anhydride (acetic, succinic, diacetyltartaric). For the most part these have the structures shown:

$CH_3(CH_2)_nCOOCH_2CH(OH)CH_2OH$ 1-monoacylglycerol

$CH_3(CH_2)_nCOOCHCH(OH)CHOCOR$ modified monoacylglycerol

where R is CH_3 (from acetate), $CH(OH)COOH$ (from lactate), CH_2CH_2COOH (from succinate), $CHOAcCHOAcCOOH$ (from diacetyltartrate), and $CH_2C(OH,COOH)CH_2COOH$ (from citrate).

Propylene glycol ($CH_3CHOHCH_2OH$) also reacts with fatty acids to give mixtures of mono (about 55%, mainly 1-acyl) and diacyl esters (about 45%). A 90% monoacyl fraction can be obtained by molecular distillation.

Other compounds include the partial esters of polyglycerols (a polyether with 2–10 glycerol units but mainly 2–4 units), sorbitan and its polyethylene oxide derivatives, 6-monoacylate sucrose, and stearoyl lactate, usually as the sodium or calcium salt.

8.12   Functional foods

There is a growing realisation of the link between diet and health. This is important to the increasing number of older people who want to live active and healthy lives as long as possible, to governments concerned about increasing health costs, and to the food industry oppressed by small margins on most conventional foods and looking for new products with higher profit margins. In the 30-year period 1995–2025 it is estimated that in the developed world the number of people over 80 years will increase by 50% and the number of those over 90 years will double. All this is linked with concerns that the food chain should deliver safer and healthier foods and to changed eating habits where less food is prepared in the domestic kitchen but is prepared in the factory for home consumption or is eaten out of the home. Under both these conditions there is some loss of control over the selection of food to be consumed. These trends are of interest to scientists and nutritionists seeking to improve the diet, to the food industry, to government, and to the media. This increasing interest is especially marked in Japan where for several years there has been a system in place for recognition of 'foods for specific health use' or FOSHU with its own distinctive logo. Many other developed countries such as Finland have also shown considerable interest in this topic.

Functional foods contain bio-active molecules that promote health or reduce the risk of disease and are consumed as conventional foods at levels not very different from normal. A recent book (Gunstone, 2003) devoted to this topic included chapters on antioxidants (carotenoids, tocopherols, flavonoids, rosemary oil, rice bran oil, and sesame oil), diacylglycerols, phytosterols, structured lipids, omega-3 acids, and CLA. Some of these have been covered in earlier sections of this chapter.

8.13   Appetite suppressants

At a time when obesity is becoming more common and is of growing concern to individuals, to the medical profession, to government health managers, and to the food industry there is increasing interest in procedures by which energy intake might be reduced (Section 7.12). There are several approaches to this issue including reducing the energy level of foods, of producing and consuming smaller portions, and through ingestion of materials that influence satiety. Satiation usually refers to the process which leads to the termination of a meal through felling 'full' and to consequent delay between meals. Several compounds are reported to enhance satiety and some are available to be added to normal foods such as soups and yoghurt.

Among the materials claimed to promote satiety are plant glycolipids, fatty acid amides, pinolenic acid ($5c,9c,12c$-18:3) present in Korean pine nut oil, and an emulsion of a mixture of purified lipid fractions from palm and oat oils. This latter, called 'olibra' and sold under various names, prolongs the feeling of fullness resulting in energy intake reduction at subsequent meals.

References and Further Reading

Bonow, R.O. and Eckel, R.E. (2003) Diet, obesity, and cardiovascular risk, *New England Journal of Medicine*, **348**, 2057–2058.

Breivik, H., editor (2007) *Long-Chain Omega-3 Specialty Oils*, The Oily Press, Bridgwater, UK.

Christie, W.W. (1998) Gas chromatography-mass spectrometry methods for structural analysis of lipids. *Lipids*, **33**, 343–353.

Christie, W.W. (2003) *Lipid Analysis*, The Oily Press, Bridgwater, UK.

Chumpitaz, L.D.A., Coutinho, L.F., and Meirelles, A.J.A. (1999) Surface tension of fatty acids and triglycerides. *J. Am. Oil Chem. Soc.*, **76**, 379–382.

Coultate, T.P. (2002) *Food – The Chemistry of Its Components*, 4th edition, The Royal Society of Chemistry, Cambridge, UK.

Coupland, J.N. and McClements, D.J. (1997) Physical properties of liquid edible oils. *J. Am. Oil Chem. Soc.*, **74**, 1559–1564.

Cuppett, S.L. (2003) Cholesterol oxides: sources and health implications, *Lipid Tecnology Newsletter*, **9**, 9–13.

Diehl, B.W.K. (2002) ^{31}P-NMR in phospholipid analysis. *Lipid Technol.*, **14**, 62–65.

Dobson, G., editor (2001) Spectroscopy and spectrometry of lipids. *Eur. J. Lipid Sci. Technol.*, **103**, 815–840; 2002, **104**, 36–68.

European Pharmacopoeia 5th edition, 2004, Strasbourg, France.

Firestone, D., editor (1999) *Physical and Chemical Characteristics of Oils, Fats, and Waxes*, AOCS Press, Champaign, USA.

Frankel, E.N. (2005) *Lipid Oxidation*, 2nd edition, The Oily Press, Bridgwater, UK.

Gunstone, F.D., editor (2003) *Lipids for Functional Foods and Nutraceuticals*, The Oily Press, Bridgwater, UK.

Gunstone, F.D. (2004) *The Chemistry of Oils and Fats*, Blackwell Publishing, Oxford, UK.

Gunstone, F.D., editor (2006) *Modifying Lipids for Use in Food*, Woodhead Publishing Limited, Cambridge, UK.

Gunstone, F.D., (2008) *Phospholipids*, The Oily Press, Bridgwater, UK (in press).

Gunstone, F.D. and Padley, B.F. (1997) *Lipid Technologies and Applications*, Marcel Dekker, New York.

Gunstone, F.D. and Herslof, B.G. (2000) *Lipid Glossary 2*, The Oily Press, Bridgwater, UK. This book can be down loaded free via The Oily Press website.

Gunstone, F.D., Harwood, J.L., and Dijkstra, A.J., editors (2007) *The Lipid Handbook with CD Rom*, 3rd edition, CRC Press, Baco Raton, FL, USA.

Gurr, M.I. (1999) *Lipids in Nutrition and Health – a Reappraisal*, The Oily Press, Bridgwater, England.

Hamilton, R.J. and Cast, J., editors (1999) *Spectral Properties of Lipids*, Sheffield Academic Press, Sheffield, UK.

Hamm, W. and Hamilton, R.J., editors (2000) *Edible Oil Processing*, Sheffield Academic Press, Sheffield, UK.

Keys, A. Andersen, J.T., and Grande, F. (1957) Prediction of serum–cholesterol responses of man to changes in the fats in the diet. *Lancet*, **2**, 959–996.

Knothe, G. (2002) Structure indices in fatty acids. How relevant is the iodine value? *J. Am. Oil Chem. Soc.*, **79**, 847–854.

Knothe, G. (2003) Quantitative analysis of fatty compounds by ^1H NMR, *Lipid Technol.*, **15**, 111–114.

Larsson, K.L., Quinn, P., Sato, K., and Tiberg, F. (2006) *Lipids: Structure, Physical Properties and Functionality*, The Oily Press, Bridgwater, UK.

Lawler, P.J. and Dimick, P.S. (2002) Crystallisation and polymorphism in fats, in *Food Lipids: Chemistry, Nutrition, and Biotechnology*, 2nd edition (C.C. Akoh and D.B. Min, editors), Dekker, New York, pp. 275–300.

Lidefelt, J.-O. (2002) *Handbook – Vegetable Oils and Fats*, Karlshamns AB, Karlshamn, Sweden.

Lonchampt, P. and Hartel, R.W. (2004) Fat bloom in chocolate and compound coatings. *Eur. J. Lipid Sci. Technol.*, **106**, 241.

McLean, B. and Drake, P. (2002) *Review of Methods for the Determination of Fat and Oil in Foodstuffs*, Review No. 37, Campden and Chorleywood Food Research Association Group, Chipping Campden, UK.

Mela, D.J. (2005) *Food, Diet, and Obesity*, Woodhead Publishing Limited, Cambridge, UK.

O'Brien, R.D. (2004) *Fats and Oils*, 2nd edition, CRC Press, Boca Raton, USA.

Rajah, K.K., editor (2002) *Fats in Food Technology*, Sheffield Academic Press, Sheffield, UK.

Ratnayake, W.M.N., Pelletier, G., Hollywood, R., Bacler, S., and Leyte, D. (1998) trans Fatty acids in Canadian margarines: Recent trends. *J. Am. Oil Chem. Soc.*, **75**, 1587.

Rossell, B., editor (1999) *LFRA Oils and Fats Handbook Series, Volume 1, Vegetable Oils and Fats*, Leatherhead Food RA, Leatherhead, UK.

Rossell, B., editor (2001) *LFRA Oils and Fats Handbook Series, Volume 2, Animal Carcase Fats*, Leatherhead Food RA, Leatherhead, UK.

Rossell, B., editor (2003) *LFRA Oils and Fats Handbook Series, Volume 3, Dairy Fats*, Leatherhead Food RA, Leatherhead, UK.

Sato, K. (2001) Crystallisation behaviour of fats and lipids – A review. *Chem. Eng. Science*, **56**, 2255–2265.

Shahidi, F. (2005) *Bailey's Industrial Oil and Fat Products*, 6th edition, 6 volumes, John Wiley and Sons, Hoboken, USA.

Sikorski, Z.E. and Kolakowska, A. (2003) *Chemical and Functional Properties of Food Lipids*, CRC Press, Boca Raton, USA.

Stier, R.F. (2004) Tests to monitor quality of deep-frying fats and oils. *Eur. J. Lipid Sci. Technol.*, **106**, 766–771.

Timms, R.E. (1978) Heats of fusion of glycerides. *Chem. Phys. Lipids*, **21**, 113–129.

Timms, R.E. (1985) Physical properties of oils and mixtures of oils. *J. Am. Oil Chem. Soc.*, **62**, 241–248.

Timms, R.E. (2003) *Confectionery Handbook*, The Oily Press, Bridgwater, UK.

Tseng, C.-H. and Wang, N. (2007) Fast analysis of fats, oils and biodiesel by FT-IR, *Lipid Technol.*, **19**, 39–40.

Warner, K. (2007) Increasing gamma- and delta-tocopherols in oils improves oxidative stability. *Lipid Technol.*, **19**, 229–231.

Whitehurst, R.J. (2004) *Emulsifiers in Food Technology*, Blackwell Publishing, Oxford, UK.

Williams, C. and Buttriss, J. (2006) *Improving the Fat Content of Foods*, Woodhead Publishing Ltd., Cambridge, UK.

Wolff, R.L., Combe, N.A., Destaillets, F., Boue, C., Precht, D., Molkentin, J., and Entressangles, B. (2000) Follow-up of the Δ4 to Δ16 *trans*-18:1 isomer profile and content in French processed foods containing partially hydrogenated vegetable oils during the period 1995–1999. Analytical and nutritional implications. *Lipids*, **35**, 815.

Useful Websites

FTIR procedures www.agrenv.mcgill.ca.foodsci.irg.irhomepg.htm

The Lipid Library maintained by Christie, W.W. www.lipid.co.uk

The USDA nutrient database www.nal.usda.gov/fnic/foodcomp

See also links www.pjbarnes.co.uk through Oily Press and through The Lipid Library to cyberlipid and SOFA.

Index

INDEX

Printed and bound by CPI Group (UK) Ltd, Croydon, CR0 4YY

27/10/2024

14580384-0001